Excel 2013 for Marketing

Workbook of Statistical Tools and Techniques for Quantitative Marketing Research and Strategy

By Tim J Smith, M. Gail Vermillion, and Myril Bruns Hillman

Wiglaf LLC
2607 W Augusta Blvd
Chicago, IL 60622

Printed in the United States of America
Printing No.: 1

ISBN-10: 0-692-40926-8
ISBN-13: 978-0-692-40926-8

TABLE OF CONTENTS

ACKNOWLEDGMENTS

Much appreciation goes to Sue Fogel and the Department of Marketing of DePaul University for providing us with the access to research materials for the development of this text, the opportunity to test the materials in the classroom, and the encouragement to do so.

INTRODUCTION

Executives need information for making rational decisions regarding marketing. Information is used to understand the firm's competitors and other marketing environments. It is used to identify and understand a target market, and make decisions regarding the proper product, promotion, placement, and pricing strategy required to attract and retain profitable customers.

Actionable information for creating and capturing profitable customers rarely comes in the form of "do this to get that". Rather, it is derived from raw data which must be aggregated and analyzed to enable a clear picture of the facts. Moreover, raw data doesn't just arrive; it has to be collected through measurements.

One of the key differences between the quantitative techniques used in marketing and those used in other fields of business, such as accounting or operations, is that marketing data comes from the market, and, more specifically, its customers in most cases. Customer data, however, is rarely as clear as a T-account in accounting or an efficiency report from operations. As such, there isn't a "single" method that can be used to interpret all marketing data. Rather, students of quantitative marketing are required to have a large number of methods and techniques at their disposal for managing and interpreting raw data into actionable information. In many ways, quantitative marketing is far more analytically demanding than most other fields in business.

In this text, we will reveal many of the methods that are used to convert raw data regarding a market into actionable information. Some of these methods require aggregating raw data and presenting it graphically so that the meaning of the data becomes readily apparent to the decision maker. Other methods require conducting statistical analysis to determine whether a data point is meaningful, and, more specifically, whether the data collected by one measurement is similar, different, or related to that collected by another measurement.

We begin this investigation into quantitative tools and techniques in marketing by examining means of interpreting single measurements on a population sample. After understanding how to manage single measurements, we will examine methods to compare the results of the different measurements on different sample populations.

Outside of Census and Economic data provided by the US government, all the data in this manual is pure fiction. The example data sets have been constructed solely for the purposes of educational instruction in applying quantitative and statistical methods to marketing challenges. No data represents the facts concerning a brand, company, market, or market segment.

This textbook has been updated to reflect Excel 2013.

CHAPTER 1: DATA TYPES & MEASUREMENT SCALES

Marketing data can be categorized into two types: Categorical or Numerical. Categorical data can be sub-divided into the subtypes of Nominal or Ordinal. Numerical data can be sub-divided into the subtypes of Interval or Ratio. Each data type and subtype must be managed differently. Hence, it is imperative that a quantitative marketer know the type of data they are working with in order to identify the proper graphical representation and analytical technique for its interpretation.

In this chapter, students will learn:

- The different types of data, including categorical (nominal & ordinal) and numerical (interval and ratio).
- The appropriate use of a pie-chart in graphical analysis of a dataset.
- The construction of a pie-chart and column-chart.
- The development of probabilities given a frequency distribution.
- The use of weighted averages for converting nominal data into ratio data.

CATEGORICAL DATA

Categorical data isn't numerical data; rather, it describes the items in a group. In effect, categorical data sorts the sample into categories. There are two subtypes of categorical data, nominal and ordinal.

NOMINAL SCALES

Categorizing items into groups requires a nominal scale. For marketing purposes, the items being categorized are often buyers and the groups often represent market segments. Because target marketing and market segmentation are required for successful market strategy, nominal scales are quite common in marketing. This enables marketers and sales executives to identify their best customers.

Nominal scales can be found with demographic, firmographic, geographic, psychographic, and behavioral segmentation to name a few examples (see table).

A graphical analysis of nominal data typically uses pie charts, bar charts, or column charts. **When and only when the categories account for 100% of the sample, pie charts can be used to show the data. In all other cases, bar, column, or some other form of chart is used.** In fact, even when the categories define 100% of the sample, both stacked bar and stacked column charts are sometimes used rather than pie charts. Selecting between the use of a pie-chart and a stacked bar or column when the categories define 100% of the sample is a subjective artistic communication choice of the analyst.

Example Nominal Scales

Race
- White
- Hispanic
- African American
- Native American
- Asian
- Hawaiian & Pacific Islander

Gender
- Female
- Male
- Unknown

Generational Cohort
- World War II
- Baby Boomer 1
- Baby Boomer 2
- Gen X
- Millennial

Population Density
- Urban
- Suburban
- Rural

US Market Region
- West Coast
- East Coast
- Central US

Global Market Regions
- Americas (North, South, and Central America)
- EMEA (Europe, Middle East, and Africa)
- APAC (Asian Pacific)

VALS Psychographics
- Innovators
- Thinkers
- Achievers
- Experiencers
- Believers
- Strivers
- Makers
- Survivors

Shopping Behavior
- Shop from a list
- Shop without a list

Category Involvement
- Purchase daily
- Purchase weekly
- Purchase monthly
- Purchase yearly
- Never purchase

Brand Choice
- Manufacturers Brand
- Store Brand
- Generic

Industry Type or NAICS code

11	Agriculture, Forestry, Fishing and Hunting
21	Mining, Quarrying, and Oil and Gas Extraction
22	Utilities
23	Construction
31-33	Manufacturing
42	Wholesale Trade
44-45	Retail Trade
48-49	Transportation and Warehousing
51	Information
52	Finance and Insurance
53	Real Estate and Rental and Leasing
54	Professional, Scientific, and Technical Services
55	Management of Companies and Enterprises
56	Administrative and Support and Waste Management and Remediation Services
61	Educational Services
62	Health Care and Social Assistance
71	Arts, Entertainment, and Recreation
72	Accommodation and Food Services
81	Other Services (except Public Administration)
92	Public Administration

Ordinal Scales

When different items within a sample can be classified in relationship to each other without fully defining the exact difference between the items, we can use an ordinal scale. **Ordinal scales define the direction of the difference, but not the exact amount of the difference.** A rank ordering of the items is the most common example of an ordinal scale.

Ordinal scales are found with lists, preferences, and other measurements.

Example Ordinal Scales		
• Ranked 1st, 2nd, 3rd, ... nth in Class • Richest Man in Latin America is Carlos Slim in 2010	• Prefer grapefruit over grapes • Prefer Chicago Style Garden Dog at $3 to a McDonalds Big Mac at $2.79	• Always, often, sometimes or rarely demand Nike shoes • Always, often, sometimes or rarely buy gum

Numerical Data

With numerical data, the data itself is measured on a numerical scale. There are two subtypes of numerical data: interval and ratio.

Interval Scales

When the items can be classified in relationship to each other and the differences between them can be associated with a meaningful quantitative measure, the items can be measured on an interval scale.

Equality in the units of difference between the points of observation in interval scales adds much clarity to a measure, even though the size of the intervals is up to the researcher. For instance, measuring items on an interval scale enables one to identify how large the difference between items is. One can claim that measure A was two units lower than Measure B, and, because the intervals are constant, researchers can use standard statistical measures of means, standard deviations, and correlation in analyzing data from measured on an interval scale.

Interval scales, however, have no meaningful zero. That is, the starting point on an interval scale could be 1, 0, -5, or wherever else the research chooses. The flexibility of choosing the starting point and interval sizes in an interval scales requires the marketing analyst to provide more clarity in their reporting of a measurement than with other scales. For instance, if we measured an item and got a 4, we don't know if the 4 is measured on a 1-10 scale, a 1-5 scale, a -5 to + 5 scale, or any other scale, and therefore we don't know if it is a low, high, or very high score. **With interval scales, the researcher must report where the scale starts and finishes.** For example, a score of 4 on a 0 to 6 point scale.

Interval scales are very common. Researchers use them with respect to preferences, sizes, category involvement, and many other measurements. For instance, Likert scales are 5 point scales used to measure the level of agreement with a proposition and are usually numerically scored starting at 1 and going up to 5. Alternatively, semantic differential scales are 7 point scales used to measure attitudes towards brands by asking people to situate a brand in relationship between two opposing adjectives. See table for other examples of interval scales.

Example Interval Scales		
Likert Scales	**Sizes**	**Semantic Differential Scales**
Please **rate your agreement** with this statement: "I love pickled jalapenos." 5. Strongly Agree 4. Agree 3. Uncertain 2. Disagree 1. Strongly Disagree **How likely are you to** purchase a Nestle Grape Soda 5. Very Likely 4. Somewhat Likely 3. Neither Likely nor Unlikely 2. Somewhat Likely 1. Very Likely	**Men's Shoe Size** (7, 8, 9, 10, 11, 12 ... yet there is no meaningful size 0 shoe) **Women's Dress size** (0 through 18, yet size zero doesn't imply that the woman doesn't exist, rather simply that she is comparatively very thin.) **Temperature if ...** Fahrenheit Celsius	**I find the Ford Focus to be** Fast _ _ _ _ _ _ _ Slow Contemporary_ _ _ _ _ _ _ Traditional Intellectual _ _ _ _ _ _ _ Emotional **I find the Marketing to be** Interesting _ _ _ _ _ _ _ Boring Rigorous _ _ _ _ _ _ _ Fluffy Intellectual _ _ _ _ _ _ _ Emotional **I find the Apple iPad to be** Hot _ _ _ _ _ _ _ Cold Hard _ _ _ _ _ _ _ Easy Intellectual _ _ _ _ _ _ _ Emotional

Ratio Scales

Ratio scales are similar to interval scales with the added feature of having a meaningful zero. Thus, ratio scales provide the same clarity, enable the same statistical measurements, and allow for the same kinds of statements regarding differences, as interval scales. **Because ratio scales have a meaningful zero, they also allow for statements such as "measure A is twice as large as Measure B".**

This distinction of having a meaningful zero is quite significant. With a ratio scale of weight measured in lbs, it might be realistic to claim that a child is half as heavy as an adult. However, with an interval scale of shoe sizes, we could not reliably say that a size 4 is half as large as a size 8.

Ratio scales are often found with respect to concrete descriptions.

Example Ratio Scales		
• Age • Height • Weight • Distance from store or manufacturer	• Units sold • Grades on a test • Stock prices • Interest rates • Income	• Business Revenue • Number of Employees • GDP per capita • Price

Graphical Analysis of Nominal Data

As stated, nominal data is often plotted with pie charts and column charts. We will start by showing how pie and column charts can be made in Excel. Excel is a highly flexible tool which allows for many ways to perform an analysis. We will describe some of them. Students may use any method they choose to complete the same analysis, but the results must be similar.

Once a chart is made in Excel, marketing analysts can "Copy & Paste" the completed graph into a word document or other Microsoft Office product.

Pie Charts

Given the data on the number of people in each race in a given area, we can create a pie chart to graphically represent the racial makeup of an area. This type of analysis might be useful for a marketer who discovers their product tends more to be purchased by a one specific racial group, or group of races, than most others.

From the US Census Bureau, using American Community Survey, the following data was collected regarding the racial demography of the Chicago Metropolitan Statistical Area for 2008.

Geographic Area: Metropolitan Statistical Area	Chicago-Naperville-Joliet, IL-IN-WI
Total population	9,502,094
RACE	
White alone	5,361,902
Hispanic or Latino (of any race)	1,849,486
Black or African American alone	1,660,131
American Indian and Alaska Native alone	10,785
Asian alone	491,477
Native Hawaiian and Other Pacific Islander alone	3,302
Some other race alone	25,182
Two or more races	99,829

We can use the following steps to make a pie chart of this data. We choose to use a pie-chart with this data because the analyst does not wish to associate a meaningful order of the races.

- Select a blank cell away from any of the data to prevent Excel from grabbing an erroneous dataset and attempting to automatically create the pie chart. It is necessary for analysis to be able to control how Excel makes plots to ensure that the correct information is communicated.
- Go to the "Insert" tab and click on "Pie" to insert a pie chart. Select the "Simple Pie" option. A blank box should appear. You may need to move the box around in the spreadsheet so that it doesn't cover up the data you wish to plot.

- In the "Chart Tools" tab grouping, select the "Design" tab, and click on "Select Data" to select the data that will be plotted. A dialogue box will appear.
    - Under the "Legend Entries (Series)", click on "Add" to add the series values of the dataset to be plotted.
        - For "Series Name", click on the small red arrow button on the spreadsheet background, and select the cell which describes the Series Name. In this case, it is the cell that reads "Chicago – Naperville – Joliet, IL-IN-WI." Hit the red return button to go back to the "Edit Series" dialogue.
        - Next, we select the series values. Click on the small red arrow button on the spreadsheet background next to "Series Values", and select the cells that describe the number of people in each race in the statistical area. In this case, it is the column of numbers starting from "5,361,094" and going down to "99,829". Hit the red return button to go back to the "Edit Series" dialogue.
        - Click "ok" after the series name and values have been selected.

At this point, your pie chart should appear. However, none of the wedges will have been labeled. If the chart isn't labeled properly, it won't communicate anything. Executives need to be able to read the chart and quickly determine its meaning. Labels enable understanding, and therefore it is always important to label graphs.

- In the "Chart Tools" tab grouping and within the "Design" tab, the "Select Data" dialogue is used to select data labels an horizontal (X) axis labels.
    - To add the labels, click "Edit" below the "Horizontal (Category) Axis Labels" in the "Select Data" dialogue. Click on the small red arrow button on the spreadsheet background next to "Axis label range", and select the cells that identify the races associated with the data. In this case, it is the column of labels starting from "White alone" and going down to "Two or more races". Hit the red return button to go back to the "Edit Series" dialogue. Click "ok" after the labels have been selected.
    - Click "ok" after the data to be plotted has been selected.
- At this point, your pie chart should appear with an accurate legend. You may want to adjust the size of the plot in order to ensure that all of the racial groups are identified in the legend. Just select a handle of the plotting box and drag it down until the legend is fully described.
- Finally, to add further clarity, we can add the sizes and portions of the different segments.
    - Click in the plotting box to bring up the "Chart Tools" tabs and go to the "Layout" tab. In the "Layout" tab, click on "Data Labels" and select "Best Fit". Now, each wedge of the pie will be labeled with the absolute size of that market segment.
    - Furthermore, click on the "Data Labels" button on the "Chart Tools – Layout" tab. Select "More Data Label Options". A dialogue box will appear for you to use to specify how the plot should appear. Include the "Percentage" check-box in the label options and hit "Close".
- At this point, your pie chart should appear fully labeled with an accurate legend. If you desire, you can play with the different plotting options such as the color in the "Design" tab. Your plot should look like that shown below.

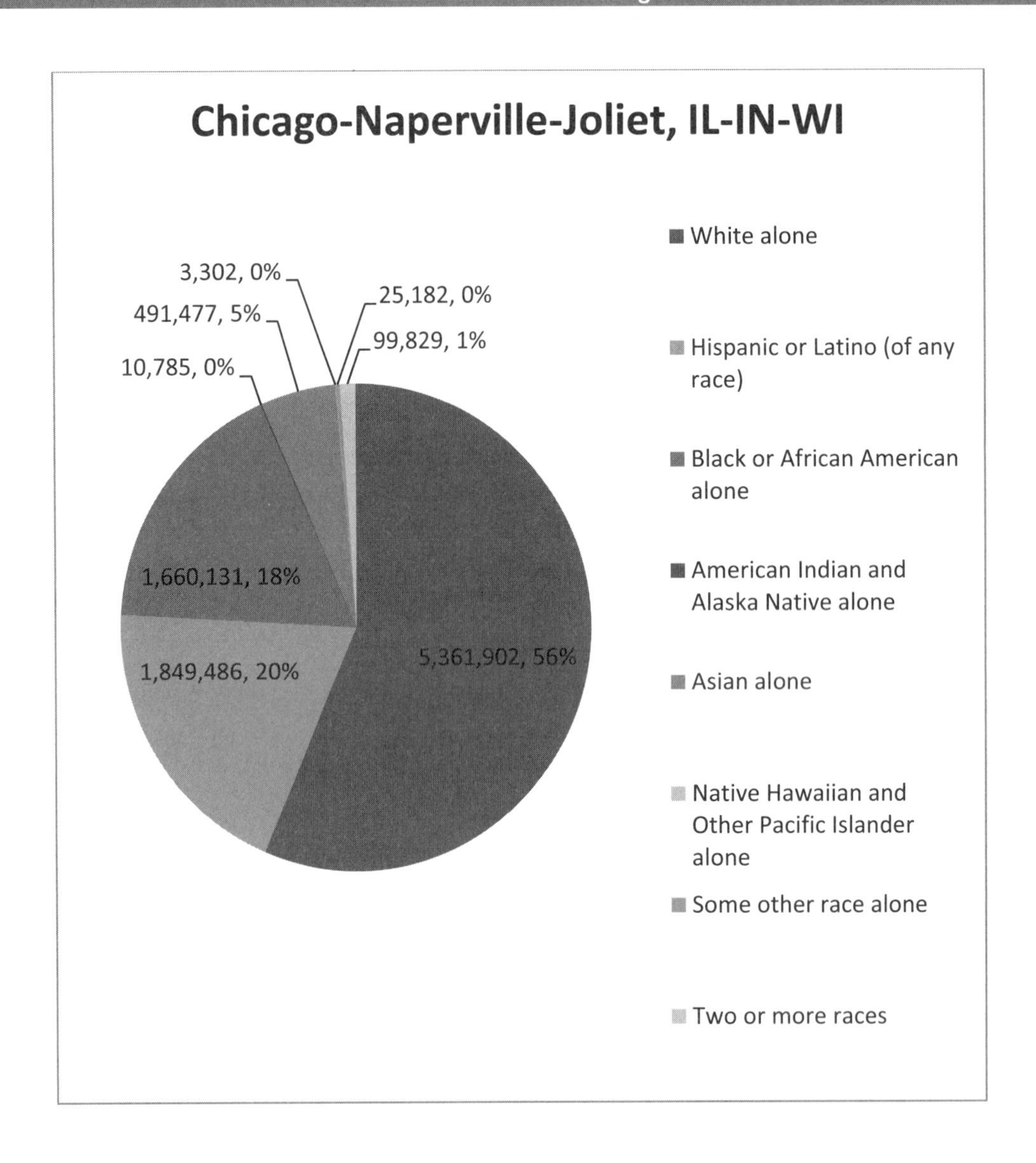

Column Charts

Given the data on the number of people in each age group in a given area, we can create a column chart to visually communicate the age profile of an area. This type of analysis might be useful for a marketer who discovers that their product tends more to be purchased by a specific age group, or group of ages, than most other ages.

From the US Census Bureau, using American Community Survey, the following data was collected regarding the age demography of the Chicago Metropolitan Statistical Area for 2008.

Geographic Area: Metropolitan Statistical Area	Chicago-Naperville-Joliet, IL-IN-WI
Total population	9,502,094
AGE	
Under 5 years	686,086
5 to 9 years	665,516
10 to 14 years	681,638
15 to 19 years	692,046
20 to 24 years	648,436
25 to 34 years	1,308,830
35 to 44 years	1,414,997
45 to 54 years	1,389,486
55 to 59 years	558,751
60 to 64 years	416,800
65 to 74 years	545,997
75 to 84 years	351,401
85 years and over	142,110

We can use the following steps to make a column chart of this data. We choose to use a column chart with this data because the analyst does wish to associate a meaningful order of the ages.

- Go to the "Insert" tab and click on "Column" to insert a column chart. Select the "Simple 2-D Column, Clustered Column" option.
- In the "Chart Tools" tab grouping, select the "Design" tab, and click on "Select Data" to select the data that will be plotted. A dialogue box will appear.
    - "Add" a series to be plotted.
        - Use the series name of the cell "Chicago – Naperville – Joliet, IL-IN-WI".
        - Use the series values of the cells that describe the number of people in each age group in the statistical area. In this case, it is the column of numbers starting from "686,086" and going down to "142,110".
        - Click "ok" after the series name and values have been selected.
    - Add the horizontal axis labels.
        - Edit the series labels by selecting the cells that identify the age categories associated with the data. In this case, it is the column of labels starting from "Under 5" and going down to "85 years and over".
    - Click "ok" after the data to be plotted has been selected.
- At this point, your column chart should appear with accurate vertical and horizontal axis labels.
- In this case, let's remove the legend. Click on the "Legend" and hit delete.

We are removing the legend in this column plot because only one data series is represented. If there were more than one dataset, such as when we were plotting the age profiles of five different metropolitan areas on the same plot, we must have a legend to identify each dataset.

- Let's add axis titles. Go to the "Layout" tab in the "Chart Tools" grouping.
    - Select "Axis Titles – Primary Horizontal Axis Title – Title Below Axis". Type in a horizontal axis title of "Age Group".
    - Select Axis Titles – Primary Vertical Axis Title – Rotated Title. Type in a vertical axis title "Frequency".
- Since the age groupings are part of a continuum and we are plotting only a single dataset on this plot, the columns should not have a gap between them.
    - To remove the gap, click on a column within the column chart, right click, and select "Format Data Series". Reduce the gap width to "No Gap".
    - Unfortunately, with no gap it can become difficult to determine when one column ends and another begins. In the same "Format Data Series" dialogue box, go to the "Border Color" tab and select a "Solid Line" of your choice.
    - Close the "Format Data Series" dialogue box.
- At this point, your column chart should appear fully labeled. If you desire, you can play with the different plotting options such as the color in the "Design" tab. Your plot should look like that shown below.

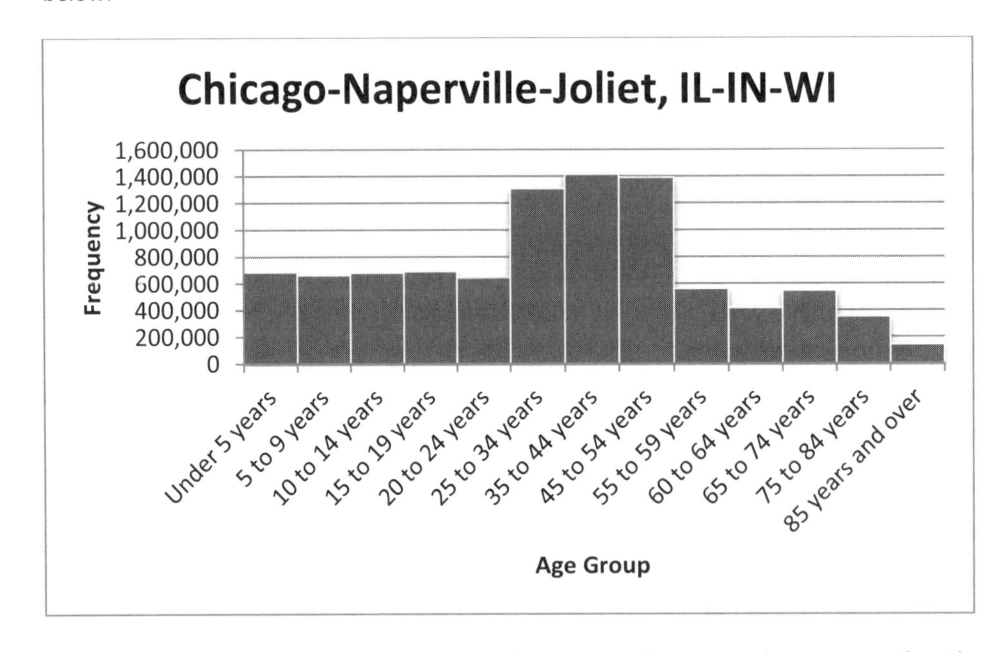

In many cases, rather than plotting the raw frequencies of a measured item, it is better to work with percentages. Raw frequencies are dependent on the population or sample size of the measure. Percentages communicate findings normalized to one (unity) and enable cross population and cross sample measurement comparisons. For instance, if we wanted to see if the Chicago area was similar to or different from the Los Angeles area in age profile, it would be better to work with percentages so that the researcher can compare the findings side by side without being confused by the overall size differences between the Chicago area population and the Los Angeles area population.

- Create a column of percentages in each age group for Chicago by dividing the frequency of an item appearing within a group by the overall population of all groups.
    - Divide the items in that age group by the total population. For instance, type "=B10/B5" in cell G10 if B10 is the cell describing the number of people Under 5 Years and B5 is the cell describing the total population size. Cell G10 will now describe the fraction of people within that age group.
    - Convert the cell's format into a percentage by clicking on the "%" button on the Home Tab.
    - It will be easy to copy and paste this cell's equation and format to the remaining cells requiring a similar calculation. But first, we must think about what we want to change for each calculation and what we must hold constant. In this case, we will need the denominator describing the total population size of a metropolitan area to be held constant as we copy the cell from row to row. To hold the cell constant from row to row, we use a dollar sign, "$", in front of the row number of the cell that should remain constant.
        - A "$" sign in front of the row number holds the cell's row constant as the cell is copied from row to row.
        - A "$" sign in front of the column letter holds the cells column constant as the cell is copied from column to column.
        - For instance, type "=B10/B$5" into cell G10 if B10 is the cell describing the number of people Under 5 Years and B5 is the cell describing the total population size. Now, B5 will be used if the formula is copied to cell G11.
    - Calculate the remaining segments' percentages by copying and pasting the formula into the required cells.
- Copy and paste the column chart of the age profile's frequency into a blank cell.
- Click on the plotted data column, and then click on the handle that captures the data series being plotted. Drag the data series selection to the percentage column.
- Change the name of the vertical axis title to "Percentage" on the new plot.
- At this point, your column chart should appear fully labeled. Your plot should look like that shown below.

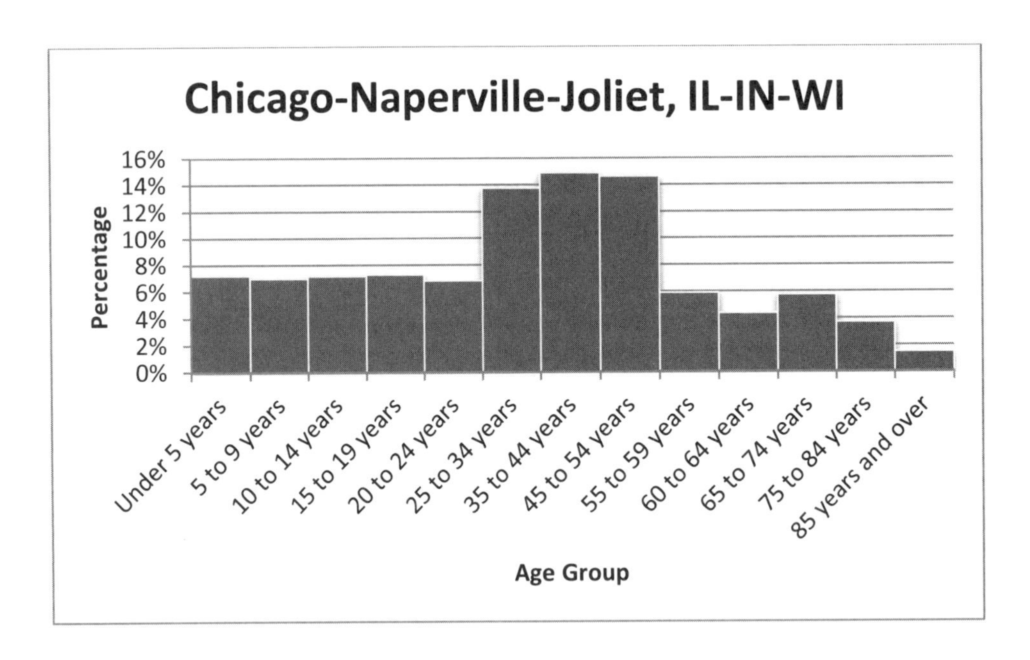

Marketing Metrics: Volume Metrics

Much of sales and marketing is focused on sales volume. Often measured by units sold, the quantity of sales has a direct impact on the firm's profitability. Finding the units sold in the past is relatively easy. Predicting the units that will sell in the future, however, can be tricky.

Sums

To find the quantity of sales in a given period, such as a week, simply count the number of units sold over that week. To find the quantity of sales over a longer period, such as six months, simply sum the units sold over every week in that six month period.

In Excel, the sum() function can be used to calculate the sum of a set of data. If the data is arranged in a column, find an empty cell below the data and label it "total". Below the data to be totaled up, type in "=sum(A1:A15)" and hit return, if the data lies in the cells A1 to A15. Use a similar argument to the sum() function for other configurations of data.

Category Incidence Frequency

To predict the units that will be sold in the future, a researcher might just assume that the past is a good predictor of the future. In many cases, this is a legitimate expectation. In some cases, such as opening a new store in a new location, there is no past to draw inferences from. In these cases, marketers must examine other forms of data.

One approach to predicting sales without directly relying on historic data is to use category incidence. Category incidence measures the frequency with which a person purchases from a product category, say, for example, pot roasts. If we know the average frequency with which pot roasts are purchased and the number of people in that neighborhood, a marketer can predict how many pot roasts will be sold in that neighborhood.

Category incidence data is usually collected using a nominal scale. Translating nominal scaled data of category incidence frequency into a meaningful numerical metric regarding the average number of pot-roasts sold per person-month requires the use of weighted averages.

Suppose a researcher wants to know, on average, how many pot roasts are sold per household in a month, and surveys households in a neighborhood with the following survey question:

How often do you eat pot roast?

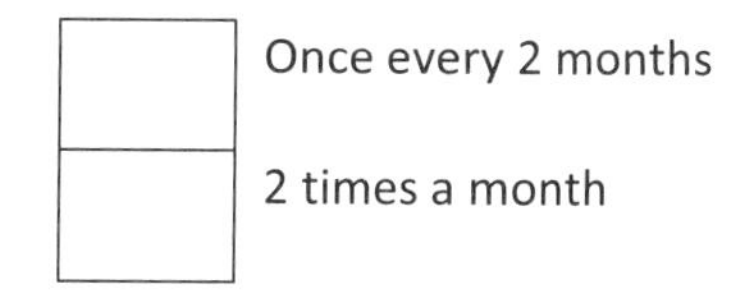

(While the possible responses to this survey question are too limited to provide meaningful information in real situations, it will suffice for the purpose of clarifying how weighted averages work).

After surveying 87 households, the researcher collects the following data on the frequency of responses:

62	Once every 2 months
25	2 times a month
87	Total Responses

To determine the expected unit sales of pot roasts per household in a given time period, the measured responses must be interpreted.

- Choose a relevant period of time to evaluate the data. In this case, we choose one month.
    - A household eating pot roast once every two months effectively eats 0.5 pot roasts a month.
    - A household eating pot roast twice every month effectively eats 2.0 pot roasts a month.
- Find the total number of purchases among the surveyed respondents in the chosen time period.
    - 62 households responded that they eat pot roast once every two months, constituting 31 sales of pot roast to this segment on a monthly basis.

$$62\ responses\ \times \frac{0.5\ sales}{response - month} = 31\ \frac{sales}{month}$$

    - 25 households responded that they eat pot roast twice a month, constituting 50 sales of pot roast to this segment on a monthly basis.

$$25\ responses\ \times \frac{2.0\ sales}{response - month} = 50 \frac{sales}{month}$$

    - Thus, the total number of sales of pot roasts to this sample per month is 81.

$$31 \frac{sales}{month} + 50 \frac{sales}{month} = 81 \frac{sales}{month}$$

- Divide by the total number of responses to determine the weighted average category incidence.
    - 81 pot roasts are sold to 87 respondents each month, making the weighted average unit sales of pot-roasts per household-month equal to 0.93 sales per household-month.

$$81\frac{sales}{month} \times \frac{1}{87\ households} = 0.93\ \frac{sales}{household - month}$$

In this brute force method, we effectively take the product of the frequency of a response and the meaning of that response, sum these products, and then divide by the sum of the responses. We could claim the same thing by saying that we are taking the sum of the products of the frequency of a response and the meaning of that response, and then dividing by the sum of the responses.

$$\frac{62\ responses \times \frac{0.5\ sales}{response - month} + 25\ responses \times \frac{2.0\ sales}{response - month}}{62\ households + 25\ housholds} = 0.93\ \frac{sales}{household - month}$$

Excel has two handy formulae that allow us to calculate the weighted average category incidence in a single calculation. As already discussed, the sum() function takes a sum of an array of values. Similarly, the sumproduct() function takes the sum of the products of the items in one array and of the items in another array.

Suppose we had the following data in Excel.

	A	B	C
1	Response Frequency	Category Incidence	
2	62	0.5	
3	25	2.0	
4			

- Typing "=sum(A2:A3)" into cell A4 will give us the number of responses from this sample, or 87.
- Typing "=sumproduct(A2:A3,B2:B3)" into cell B4 will give us the sum of the products of (62 X 0.5) and (25 X 2.0), or 81, the expected number of pot roasts sold to this sample of respondents in a given month.
- Typing "=B4/A4" into cell C4 will give us the weighted average category incidence, or 0.93 pot roasts per household-month.
- Similar, doing all of this in one step in cell C3, we could type "=sumproduct(A2:A3,B2:B3)/ sum(A2:A3)" to find the weighted average category incidence of 0.93.

Weighted average category incidence can also be calculated with response percentage data rather than response frequency data. In many cases, researchers do not have response frequency data, only response percentage data. Moreover, in comparing different measures, it is often easier to work with percentage data rather than raw frequency data as percentage data does not depend on the overall size of the data set or sample, while frequency data does.

Throughout this course, students will be required to convert frequency data into percentage data. Becoming comfortable with this process early in the course will make the learning experience easier. Just remember, divide

the frequency of an item by the total number of responses on all items. Using the same dataset, the following procedure would be used.

- Typing "=A2/sum(A$2:A$3)" into cell C2 will provide the percentage of people who state that they eat a pot roast every other month.
- Copying this formula into cell C3 will provide the percentage of people who state they eat a pot roast twice a month.
- Typing "=sumproduct(B2:B3,C2:C3)", or the sum of the products of the category incidents and their relative probability weightings, we find the weighted average category incident in cell C4.
- After managing the formats of the cells, the spreadsheet should look as follows:

	A	B	C
1	Response Frequency	Category Incidence	Response Frequency
2	62	0.5	71%
3	25	2.0	29%
4			0.93

If a butcher is considering opening a store in a neighborhood with 20,000 household, they should expect this neighborhood to purchase 18,000 pot roasts per month. The next question that butcher might ask is: "How many of those sales will I be able to capture?"

EXERCISES

METROPOLITAN AREA DEMOGRAPHICS

The data was collected for several metropolitan areas from the 2006-2008 American Community Survey 3-Year Estimates Data Set.

1. Use Excel to make a pie chart of the racial profile of each metropolitan area (5 charts in total).
   a. Title each graph with the name of the metropolitan area.
   b. Label each segment with the name of the segment size, the percentage of people within that segment, and the absolute size of that segment.
2. What kind of data are we associating with an individual when categorizing them by race or ethnicity: Nominal, Ordinal, Interval, or Ratio?
3. Which of the metropolitan areas ranks highest in percentage Hispanic? What kind of data is ranking: Nominal, Ordinal, Interval, or Ratio??
4. Which of the metropolitan areas has the highest number of Blacks or African Americans? What kind of data is a count of the number of people of a given race: Nominal, Ordinal, Interval, or Ratio?
5. Which of the metropolitan areas, Atlanta, Boston, San Antonio, or Los Angeles, has the nearest racial profile to Chicago. Why do you say this?
6. Use an Excel column chart to make column chart of the percentage of people within a given age segment of all metropolitan areas on a single plot. (You will need to calculate the percentages in each segment for each metropolitan area, and add multiple series to a single plot).
   a. Title the graph "Metropolitan Age Profile".
   b. Label the vertical axis "Percentage" and the horizontal axis "Age".
   c. Use the legend to denote the metropolitan area.
7. What age groups are overrepresented in high-density urban areas? (An overrepresented age group would be one which has a higher than expected percentage within that age group.)
8. Which city has highest percentage of individuals 60 and over?
9. Which city has the highest percentage of individuals 19 and younger?

Geographic Area: Metropolitan Statistical Area	Chicago-Naperville-Joliet, IL-IN-WI	Atlanta-Sandy Springs-Marietta, GA	Los Angeles-Long Beach-Santa Ana, CA	San Antonio, TX	Boston-Cambridge-Quincy, MA-NH
Total population	9,502,094	5,251,899	12,818,132	1,982,788	4,494,144
SEX					
Male	4,677,996	2,594,176	6,362,881	971,337	2,185,051
Female	4,824,098	2,657,723	6,455,251	1,011,451	2,309,093
AGE					
Under 5 years	686,086	411,804	927,753	159,937	271,489
5 to 9 years	665,516	389,309	878,885	147,844	266,364
10 to 14 years	681,638	382,937	950,244	151,913	285,685
15 to 19 years	692,046	362,555	980,601	146,981	312,753
20 to 24 years	648,436	330,120	924,828	146,579	316,121
25 to 34 years	1,308,830	767,882	1,779,876	280,985	574,553
35 to 44 years	1,414,997	881,301	1,992,593	278,030	699,761
45 to 54 years	1,389,486	767,917	1,792,470	264,331	695,351
55 to 59 years	558,751	306,771	694,834	108,347	277,871
60 to 64 years	416,800	223,045	533,337	84,869	226,815
65 to 74 years	545,997	250,652	711,346	111,290	283,700
75 to 84 years	351,401	128,707	466,968	73,523	195,013
85 years and over	142,110	48,899	184,397	28,159	88,668
RACE					
White alone	5,361,902	2,841,232	4,239,908	749,714	3,494,598
Hispanic or Latino (of any race)	1,849,486	487,984	5,645,374	1,047,746	354,702
Black or African American alone	1,660,131	1,614,618	892,344	118,080	281,567
American Indian and Alaska Native alone	10,785	10,265	32,291	6,569	6,145
Asian alone	491,477	216,267	1,731,576	34,271	260,607
Native Hawaiian and Other Pacific Islander alone	3,302	2,636	33,045	1,877	1,415
Some other race alone	25,182	20,508	46,575	2,825	38,682
Two or more races	99,829	58,389	197,019	21,706	56,428

FIRM PERFORMANCE

Sales performance data over 26 weeks for a firm were collected.

Week	Unit Sales	Week	Unit Sales
1	10,662	14	14,180
2	15,119	15	15,398
3	16,751	16	19,882
4	13,493	17	15,951
5	9,573	18	14,732
6	14,284	19	28,056
7	15,244	20	15,375
8	22,138	21	16,843
9	34,342	22	12,795
10	18,446	23	13,131
11	11,544	24	11,004
12	11,940	25	14,206
13	10,162	26	10,440

1. Find the total number of unit sales over the period. (Hint, use the sum() function).
2. What kind of data is unit sales: Nominal, Ordinal, Interval, or Ratio?

PASTA

A researcher wants to know, on average, how often pasta is consumed per household in a week, and surveys households in a neighborhood with the following survey question:

How often do you eat pasta? (Pick the closest figure).

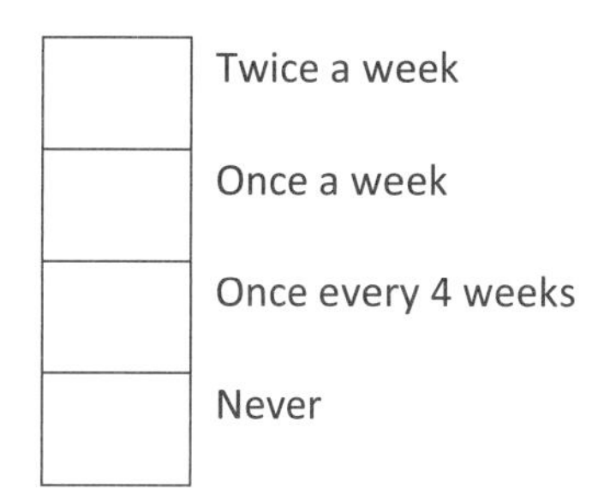

After conducting the survey, the research produces the following data:

Response Item	Response Frequency
Twice a week	14
Once a week	27
Once every 4 weeks	50
Never	7

1. What kind of data was collected: Nominal, Ordinal, Interval, or Ratio?
2. How many responses were collected?
3. What percentage of people responded in each category?
4. What is the weighted average category incidence for pasta of this sample?

Split Pea Soup

A researcher wants to know, on average, how often split pea soup is consumed per household in a year and surveys households in a neighborhood with the following survey question:

How often do you eat split pea soup? (Pick the closest figure).

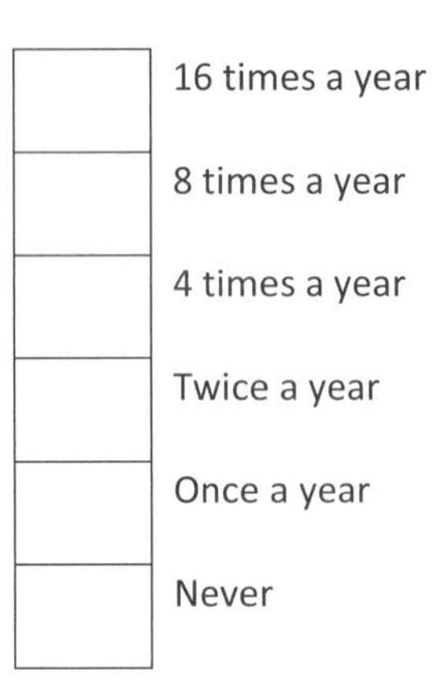

After conducting the survey, the research produces the following data:

Response Item	Response Frequency
16 times a year	17
8 times a year	23
4 times a year	52
Twice a year	186
Once a year	95
Never	137

1. How many responses were collected?
2. What percentage of people responded in each category?
3. What is the weighted average category incidence for split pea soup of this sample?

CHICKEN

A researcher wants to know, on average, how often whole chickens are purchased per household in a year, and surveys households in a neighborhood with the following survey question:

How often do you buy a chicken? (Pick the closest figure).

- ☐ Once a week
- ☐ Once every other week
- ☐ Once every 4 weeks
- ☐ Never

After conducting the survey, the research collects the following data:

Response Item	Response Probability
Once a week	7%
Once every other week	17%
Once every 4 weeks	34%
Never	42%

1. What is the weighted average category incidence for chicken of this sample?

CHAPTER 2: FREQUENCY DISTRIBUTIONS

Frequency distributions describe how often a particular item was identified among the possible items, and can be found with many different kinds of data. For instance, with market survey data, the item may be a response to a particular question; with market segmentation data, the item might be a specific market segment among the potential market segments; and with economic data, the item might the number of businesses with revenue falling within a specific range.

We have already seen a couple of frequency distributions. The metropolitan area census data on gender, age, and race provided frequency distributions for people matching a specific categorization. Similarly, the category incidence data provided the frequency distributions of responses to a survey question.

For many datasets, graphical analysis requires the construction of frequency distribution plots or column charts. **As stated, pie charts can be used with nominal data if and only if the categories account for 100% of the sample.** In all other cases, a different form of graphical analysis is required. Often, the appropriate graphical analysis requires a column or bar chart.

In this chapter, students will learn:

- How frequency distributions can be reframed through the proper use of intervals.
- The use of the Histogram dialogue in Excel to create frequency distributions.
- How to identify the skewness of a frequency distribution, including positive, negative and relatively non-skewed data.
- How to identify the kurtosis of a frequency distribution, including platykurtic, mesokurtic, and leptokurtic distributions.
- How to identify the central tendency of a distribution, including mean, median, and mode.
- How to calculate the standard deviation of a distribution.
- The Profit Equation of the Firm, including revenue and cost metrics.

INTERVALS

Frequency distributions were introduced in Chapter 1 with a variety of datasets. In those distributions, the interval size was provided in the dataset. However, the interval sizes provided by a dataset are not always the best interval size to use in representing the data. When a marketer is designing a research study, one of the key decisions to make is the size of the interval.

The size of the interval determines how the data is graphically presented. It also determines the granularity of inspection that can be conducted once the research is complete and a report has been tendered. While it might be tempting to always use the smallest level of granularity in analyzing a dataset, such a bias is unfounded. Too much granularity prevents executives from being able to quickly grasp the meaning of an analysis.

Executives may feel that they "can't see the forest for the trees" when there is excess granularity. Alternatively, they may not be able to tell what they are looking at when there is insufficient granularity. As such, the analysis must make tradeoffs between details and clarity when analyzing a dataset.

In this section, we will demonstrate how the choice of the interval size can alter the clarity of the information revealed in a table or graphical analysis.

Unit Intervals

To refresh our knowledge of frequency distributions, let us consider the distribution of women's shoe sizes. 428 females were surveyed regarding their shoe size and the following interval data was collected, where sizes were measured using whole numbers only. From the frequency data, the researcher created a frequency plot, or a plot of the frequency with which a woman from this sample wears a particular size.

Female Shoe Size	Response Frequency
Unknown	6
4	11
5	6
6	75
7	119
8	111
9	66
10	33
11	0
12	1
Total	428

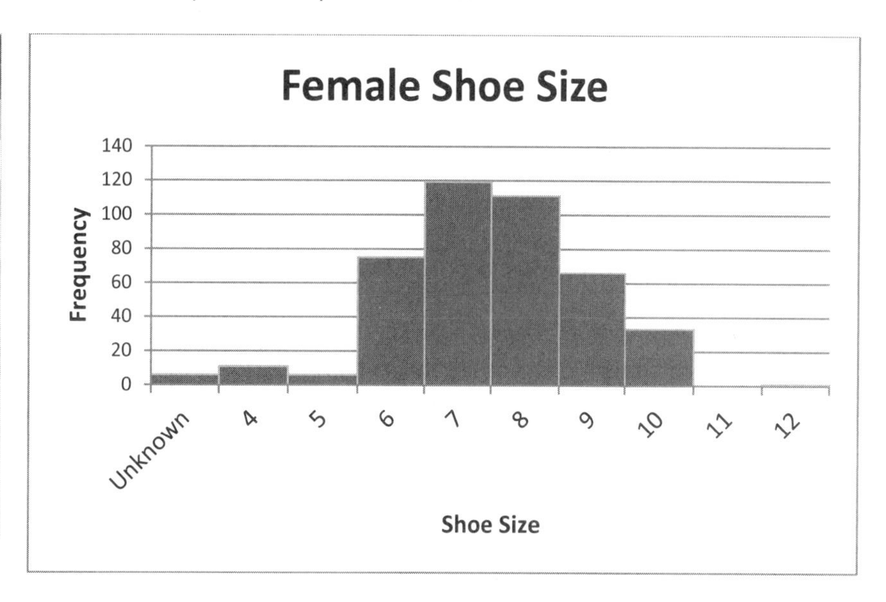

The researcher also chose to represent the data using percentages rather than raw frequencies to enable comparisons independent of sample sizes. She divided the frequency of each response items by the total number of responses collected to calculate the percentages in each category.

Female Shoe Size	Response Percentage
Unknown	1.4%
4	2.6%
5	1.4%
6	17.5%
7	27.8%
8	25.9%
9	15.4%
10	7.7%
11	0.0%
12	0.2%

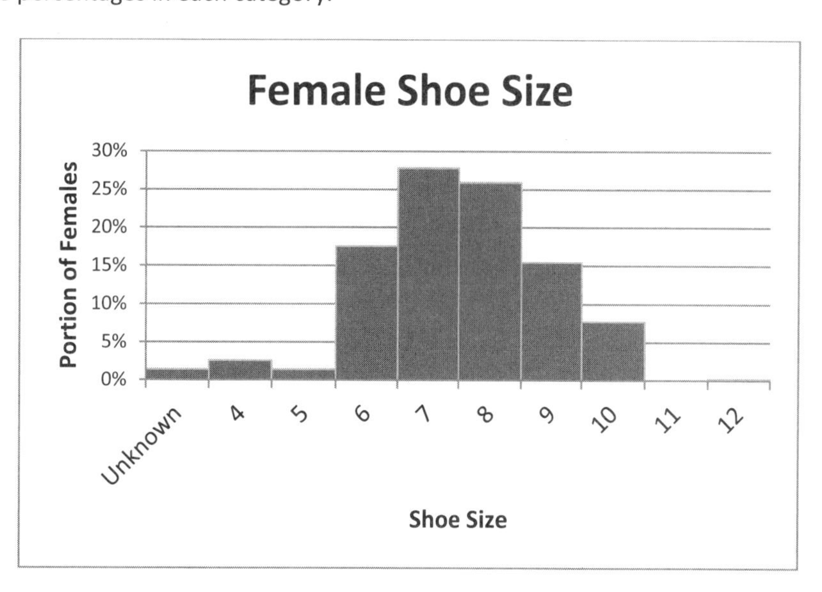

From the tables and charts, we can readily see that the maximum shoe size found in this sample is 12, and the minimum shoe size found in this sample is 4. We can also see that a small portion (1.4%) of women in this sample don't know their shoe size.

NON-UNIT INTERVALS

With some numerical data, unit intervals are too small. Rather than taking each item as its own category, we might group the items together into a broader category.

For a concrete example, let us continue with the women's shoe size dataset. An athletic gear retailer may not only need to know how many women fit each size of shoe, but also how many small, medium, and large socks needed to complement the shoe selection. The sock supplier may have told the retailer that a small sock fits women whose shoe sizes lies in the range of 4-6, a medium sock fits women whose shoe sizes lies in the range of 7-9, and a large sock fits women whose shoe sizes lies in the range of 10-12. Given this information, the retailer can calculate the required percentage of small, medium, and large socks to stock.

In this case, the interval size is 3. Small socks fit sizes 4, 5, and 6. Medium socks fit sizes 7, 8, and 9. And large socks fit sizes 10, 11, and 12.

Using the same data collected on women's shoe sizes, the retail analyst can now calculate the appropriate percentage of sock sizes. The analyst will ignore the "unknown" sizes in this analysis as this data-point provides no information useful for decision making. After conducting the calculation, the analyst will once again plot it using a column chart. From this calculation, the retailer now knows that about 22% of the female athletic socks should be smalls, 70% should be mediums, and 8% should be large. Hence, the proper S:M:L ratio of socks is 22:70:8.

See chart and graph on the next page.

Female Shoe Size	Response Frequency	Sock Size	Equivalent Shoe Size	Sock Size Frequency	Sock Size Percentage
Unknown	6				
4	11				
5	6	Small	4-6	92	22%
6	75				
7	119				
8	111	Medium	7-9	296	70%
9	66				
10	33				
11	0	Large	10-12	34	8%
12	1				

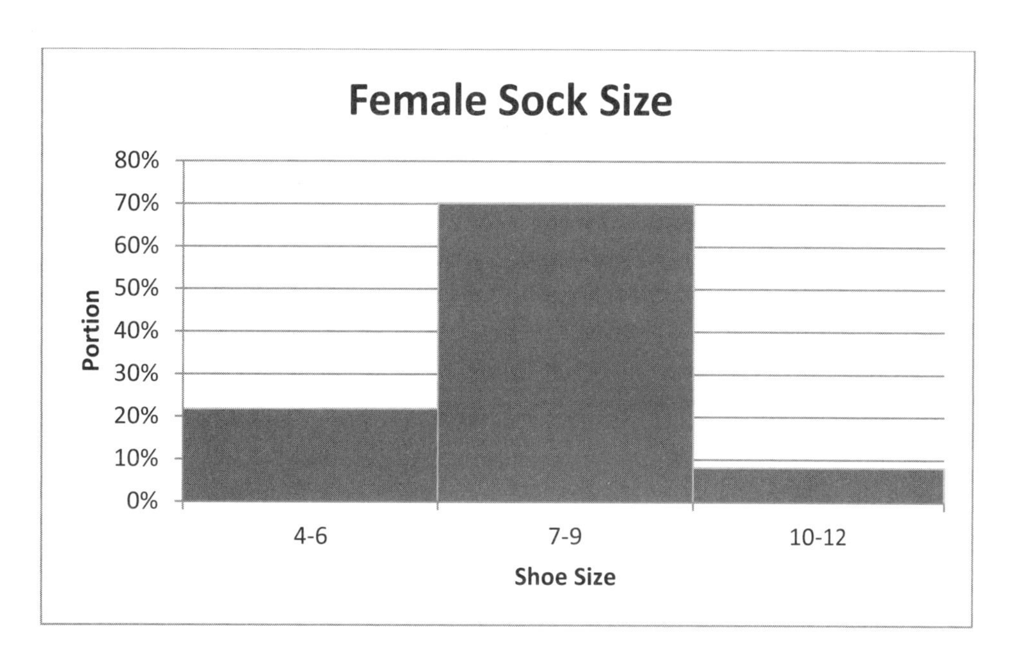

Histograms in Excel

When raw data is collected, it doesn't arrive tabulated into a nice frequency distribution. Rather, an analyst must construct the frequency analysis. While it is possible to define the intervals and count how many responses lie within each interval with pencil and paper, such an effort is tedious and error prone. Instead, data analysts can use a tool to automate some of the work.

Excel provides a histogram tool for creating frequency distributions. A histogram is a graphical representation (usually a column chart) that organizes a group of data points into user-specified ranges. To demonstrate how the histogram dialogue works in Excel, let us consider a dataset that contains the distances traveled by students to DePaul University.

A research asks students "How far did you travel to the DePaul Loop Campus today?" and collects the following raw ratio data.

Distance Traveled (miles)			
3.5	10.8	0.5	3.5
3.2	22.4	19.6	3.1
18.7	0.5	3.6	4.3
2.1	1.5	28.3	34.1
0.9	7.3	3.1	3.4
14.3	6.4	3.5	7
4.3	24.6	2.8	4
8.2	15.2	4.2	

Giving each score its own frequency measurement will create a relatively meaningless set of information. Rather, it is necessary to group the scores together and measure the number of students who traveled within given range of distances. With Excel, grouping data-points into ranges requires defining intervals and using the histogram dialogue.

First, define the intervals. The intervals must span the entire range of possible data points. They can go beyond the actual range of data, but they cannot span a range of points that are actually smaller than that contained in the dataset.

- In Excel, we can use the functions MIN() and MAX() to define the range of data collected. Label cells in blank column near the data "Min" and "Max". (Everything needs a label in order to be identified and communicated).
    - In the cell next to "Min", type in "=min(array)" where "array" identifies the cells that contain the data, perhaps from A2:D9. You should find the minimum data-point at 0.5 miles.
    - In the cell next to "Max", type in "=max(array)" where "array" identifies the cells that contain the data, perhaps from A2:D9. You should find the maximum data-point at 34.1 miles.
- Next, define the intervals. If every mile was an interval, we would require 35 different intervals. (Before we show how to define and use a smaller number of intervals, let us demonstrate how the histogram dialogue works). Label a blank column, say the "J" column, with the title "Upper Limit".
    - In the first cell down, say J2, type "1".
    - In the next cell down, type "=1+J2", or add 1 to the cell value above it.
    - Copy and paste this cell downward until you have 35 intervals. You have now defined 35 intervals of equal size. The first covers the range up to 1 mile. The second covers the range above 1 mile and up to 2 miles. The third covers the range above 2 miles and up to 3 miles. Similar statements can be made for all intervals.
- If you haven't already, add the Data Analysis ToolPak. (Use the sequence of menus found under File->Options->Add-ins-> Manage Excel Add-ins, hit "GO"-> Select "Analysis ToolPak" ->"OK")
    - Go to the "Data" tab and hit the "Data Analysis" button.
    - Select "Histogram" and hit "OK".
    - The Input Range in the Histogram dialogue requires the Excel array which contains the data to be analyzed. Hit the red arrow button on the spreadsheet background and select the array of distance traveled data to be analyzed. Then hit the red return button to return to the dialogue box.
    - The Bin Range in the Histogram dialogue requires the Excel array which contains the upper limits of the intervals to be used in analyzing the data. Hit the red arrow button on the spreadsheet background and select the column of upper limits that were created in the prior effort. Do not include the column title in your selection. (Shortcut key note: hit Ctrl-Shift-Down Arrow to quickly go to the end of an array while selecting the entire array.) Then hit the red return button to return to the dialogue box.
    - Under "Output Options", select the radial button next to "Output Range". Then, hit the red arrow button on the spreadsheet background and select a single cell that will hold the first element of the output. We suggest you put the output next to the column of upper limits of the intervals, and as such that you select the cell right next to the column title "Upper Limit". Then hit the red return button to return to the dialogue box. Failure to define where the output goes will result in the output going to a new worksheet and forcing you to go back and forth between worksheets to see both the analysis and the raw data – an unpleasant affair.

- Hit "ok".
- The output will include two columns. The first, titled "Bin", contains the position of the upper limits. The second, titled "Frequency", contains the frequency of response lying above the prior interval and up to the upper limit defined in the bin column.

- You have now created a frequency distribution given specific raw data. Next, plot it and you should get the following result. (Review the section on Column Charts in Chapter 1 if needed. Every axis should be labeled and the chart should be titled).

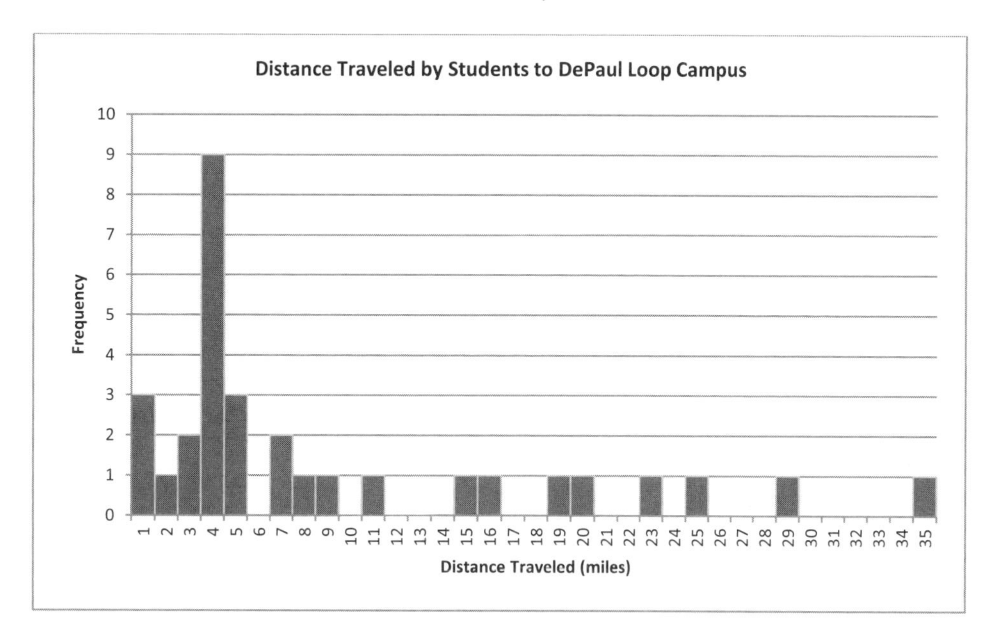

From the above chart, we see that many intervals contain no responses, others have just one, and then a few contain the majority of the responses. A frequency distribution that contains mostly ones and zeros and then a few highly popular items is an example of excess granularity that prevents a simple picture of the data. In these cases, we should try using fewer intervals. For this dataset, we will demonstrate how the data can be plotted with just 7 intervals.

- Divide the range by the desired number of intervals to get a rough estimate of the proper interval size.
  - The data goes from 0.5, or roughly 0, up to 34.1, for a range of 33.6.
  - 33.6 / 7 = 4.8
- Round this rough estimate up to a higher round number that will be used to define the intervals.
  - The intervals will be of size 5.
- Define the lower limit of each interval at a convenient round number below the minimum point in the dataset.
  - The starting lower limit is 0.
- Title blank column "Lower Limit" and the blank column next to it "Upper Limit".
- The Lower Limit column will contain the lower limits of each interval.
  - In the cell right below the title "Lower Limit", type in the starting point for the lower limits, 0. (If the Lower Limit column starts in cell I1, cell I2 should be equal to 0).
  - In the cell below or at the starting lower limit, add the determined interval size to the cell above it. (If cell I2 contains the starting lower limit, cell I3 should contain the formula "=J2+5").

  - Copy and paste the cell containing the formula for calculating the lower limits until all intervals have been calculated.
- The Upper Limit column will contain the upper limits of each interval.
  - In the cell immediately below the title "Upper Limit", add the interval size to the lower limit minus a small round-off value.
  - In this case, a natural round-off value would be 0.5. A person traveling 4.6 miles is more like a person who travels 5 miles than one who travels 4 miles.
  - (If cell I2 contains the starting lower limit, cell J2 should contain the formula "=I2+5-.05".)
  - Copy and paste the cell containing the formula for calculating the upper limits until all intervals have been calculated.

At this point, your spreadsheet might look like:

	A	B	C	D	E	F	G	H	I	J
1	Distance Traveled (miles)								LL	UL
2	3.5	10.8	0.5	3.5		Min	0.5		0	4.5
3	3.2	22.4	19.6	3.1		Max	34.1		5	9.5
4	18.7	0.5	3.6	4.3		Range	33.6		10	14.5
5	2.1	1.5	28.3	34.1		Intervals	7		15	19.5
6	0.9	7.3	3.1	3.4		Rough Interval Size	4.8		20	24.5
7	14.3	6.4	3.5	7		Used Interval Size	5		25	29.5
8	4.3	24.6	2.8	4					30	34.5
9	8.2	15.2	4.2							

We can now create a histogram of the data using the newly defined upper limits of the intervals. Using the same Histogram dialogue, selecting the same input range but this time selecting the Bin Range to be the upper limits of the 7 intervals, we would get the following output:

L	M
Bin	*Frequency*
4.5	18
9.5	4
14.5	2
19.5	2
24.5	2
29.5	2
34.5	1
More	0

- We will want to plot this frequency data with a column chart with appropriate axis labels. However, we are still lacking an appropriate column containing the horizontal axis labels. Hence, we must construct one. We have two options, either hand type each label or use a formula to automatically create them.
    - To create the horizontal axis labels with a formula, use the function in Excel that allows us to add text. The "&" function in excel joins two strings of text. Conceptually, we will type in "=LL&" – "&UL" where LL is defined by the cell with the lower limit of that interval and UL is defined by the cell with the upper limit.
    - If the columns containing the Lower and Upper Limits are in the I and J column respectively, and column K is empty, title the K column "Labels".
    - In cell K2, type in the formula "=I2&"-"&J2".
    - Copy and paste the cell containing the formula for calculating the horizontal axis labels until all have been calculated.
- We will also want to ensure that we plot percentages rather than sample frequencies to enable comparisons across populations of unequal sizes.
    - After calculating the percentages of responses within each category (refer back to the Column Charts section of Chapter 1 for details on the approach), you should get the following output and plot.

	I	J	K	L	M	K
1	LL	UL	Labels	*Bin*	*Frequency*	Percentage
2	0	4.5	0-4.5	4.5	18	53%
3	5	9.5	5-9.5	9.5	4	13%
4	10	14.5	10-14.5	14.5	2	6%
5	15	19.5	15-19.5	19.5	2	6%
6	20	24.5	20-24.5	24.5	2	6%
7	25	29.5	25-29.5	29.5	2	6%
8	30	34.5	30-34.5	34.5	1	3%
9				More	0	

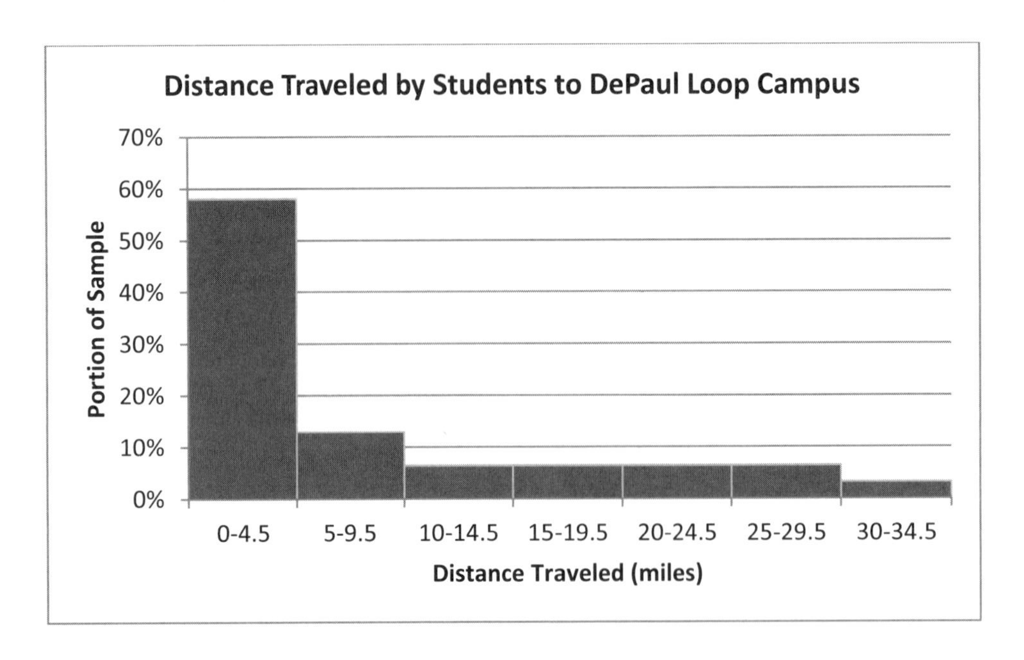

Skewness

While the frequency plot of women's shoe sizes had a clear peak at size 7 to 8 and trailed off relatively symmetrically as we considered smaller or larger shoe sizes, the distance students traveled to school appears to peak at a short distance but trail off with a long tail towards longer distances.

When a frequency distribution is clustered around a set of points and trails of in one dominant direction, that frequency distribution is skewed. More specifically,

- If the frequency distribution is overrepresented with items above the most common score, that frequency distribution is positively skewed.
- If the frequency distribution is overrepresented with items below the most common score, that frequency distribution is negatively skewed.

For instance, consider the bowling scores of three teams after 50 games on the next page. Those for the Green Tomatoes tend to trail off toward higher scores, and thus the distribution of scores is positively skewed. Meanwhile, those for the Red Cabbage Heads tend to trail off toward lower scores; we would say that their distribution of scores is negatively skewed. And the Young Turks score distribution is relatively symmetrical, hence it is not skewed. Though computational techniques can be used, identifying skewness is best done visually in most cases.

The skew of a distribution is sometimes driven by outliers in the dataset. **Outliers are data points that are extremely above or below the other data points within the dataset.**

POSITIVELY SKEWED DISTRIBUTION

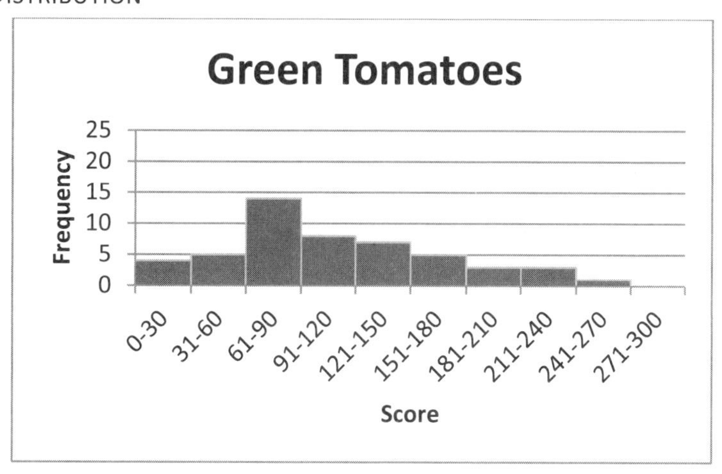

NEGATIVELY SKEWED DISTRIBUTION

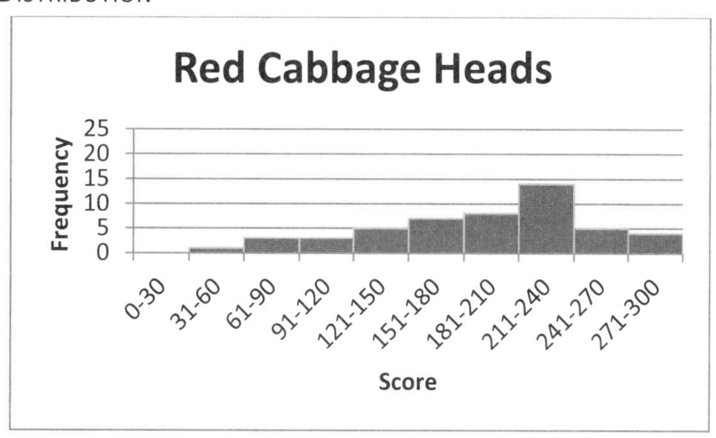

RELATIVELY UN-SKEWED DISTRIBUTION

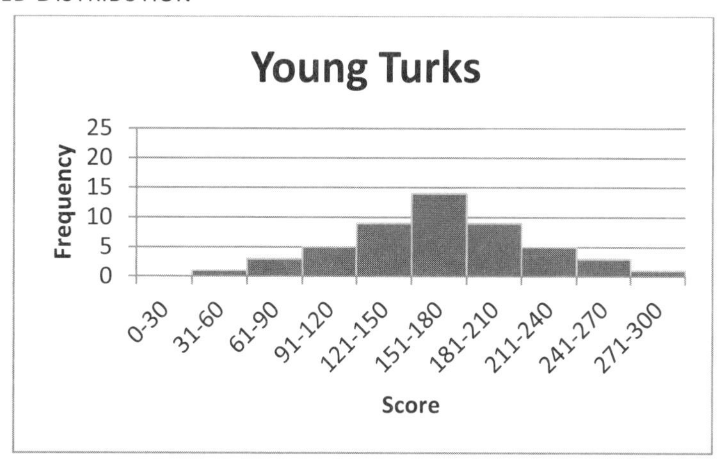

Kurtosis

When investigating a distribution, we should also examine the peakedness. Highly peaked distributions, such as that for the distance students travel, imply that a large percentage of responses fall within a single category. Broad distributions, such as that for women's shoe sizes, imply that the responses fall within a broader number of categories. Generally flat distributions, such as that for the age profile of a metropolitan area, imply that the responses are equally spread out.

Kurtosis describes the peakedness of the distribution. **Kurtosis comes in three different types, leptokurtic, platykurtic, and mesokurtic.**

- If the distribution is highly peaked, it is leptokurtic. Leptokurtic distributions have most of the responses within a single category or small cluster of categories.
- If the distribution is relatively flat, it is platykurtic. Platykurtic distributions imply that the responses are relatively equally spread across all response categories.
- If the distribution is somewhere in between, it is mesokurtic. Mesokurtic distributions imply that the responses are centered on a common category, and decline in frequency as higher and lower categories are considered. Mesokurtic distributions can be well described as a bell curve distribution.

As a mnemonic, leptokurtic distributions "leap up", platykurtic are flat like the bill on a platypus, and mesokurtic distributions lie in the middle – you know, meso.

See the next three charts for examples of leptokurtic, platykurtic, and meskurtic distributions.

Leptokurtic Distribution

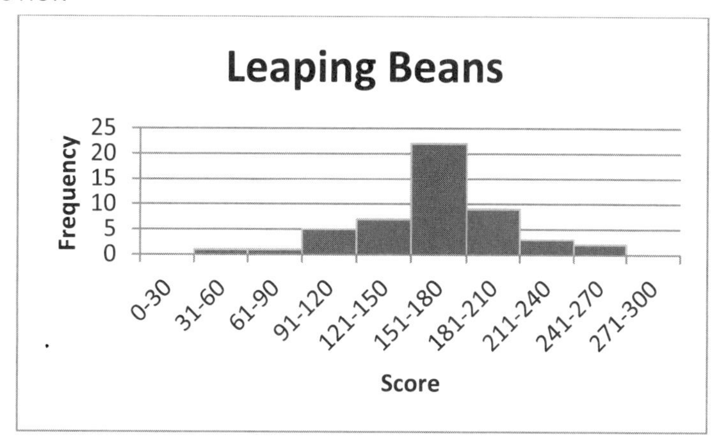

Platykurtic Distribution

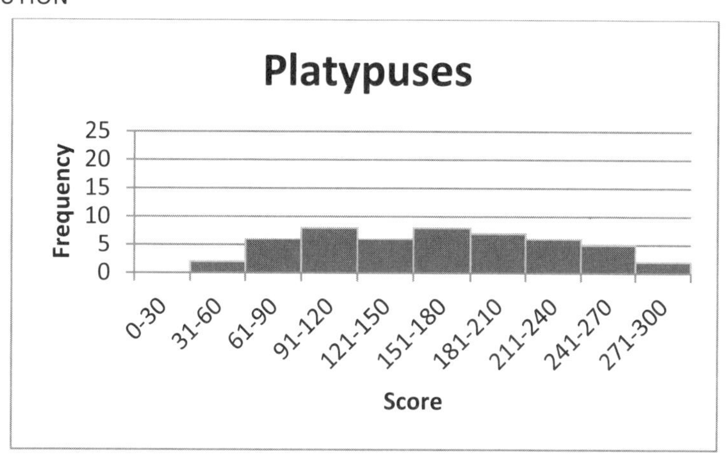

Mesokurtic Distribution

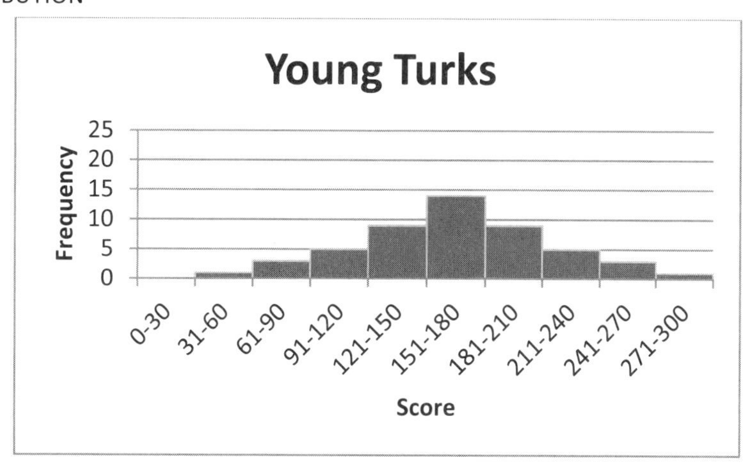

Central Tendency

The central tendency of a distribution is the single value that best describes the sample's distribution as a whole. For instance, we look at the Young Turks bowling score distribution shown above and state that the average bowling score is between 151 and 180 for this team.

There are three dominant ways to describe the central tendency of a distribution: mean, median, and mode.

Mean is the arithmetic mean, the common average of a distribution given by

$$\bar{X} = \frac{\sum_{i=1}^{N} X_i}{N}$$

where we read $\bar{X}$ as "X Bar" and Xi as "X sub i". N is the number of items in the sample and the Greek letter Σ, or sigma, means to take a sum. Hence, we would state that "X Bar equals the sum of X sub i as i goes from 1 to N, divided by N". The arithmetic mean can be found with any numerical data.

Median is the score in the middle. (Think highway median, or median means middle). 50% of a sample will give a response at or below the median, and 50% will give a response at or above the median. The median can be found with any numerical data, as well as ordinal data if the complete list or list length is provided.

Mode is the most popular score, or the score most often returned. (Think "what is in mode, fashion, or vogue?"). The mode can be found with any numerical data, as well as some nominal data if one category is preferred over all others.

For most distributions, the mean is the best representation of the central tendency of the measurement. It is also the most cited measurement of central tendency that we hear on the news. For instance, the mean shoe size shown above would be 7.5 (we discard the remaining decimals to keep the measurement of the central tendency at a similar number of significant digits as the accuracy of the measurement).

However, the mean is not always the most accurate metric of the central tendency. When the data is highly skewed, the mean will give a number that does not represent the sample.

For example, in the dataset for distance traveled to school, we would calculate that the mean distance traveled is 8.7 miles. However, just by looking at the data, we know that most people didn't travel 8.7 miles, they traveled less. In fact, half the people traveled 4.2 miles or less, and half of them traveled 4.2 miles or more. Hence, 4.2 is the median distance traveled and is a better representation of the central tendency.

As for the most popular distance traveled, we find that it is 3.5 miles. As a metric of the central tendency, this simply implies that, when asked, the most popular response was three and half miles.

With positively skewed data, the mean will be above the median. With negatively skewed data, the mean will be below the median.

Excel has handy tools for these metrics as well. Use the SUM() function to find the sum of the values in a dataset, and the COUNT() function to count the number of items in the dataset, and then calculate the mean of a dataset by using "=SUM()/COUNT()", or one could use the AVERAGE() function. (We are using () to imply that the appropriate cell, cells, or array of cells is used to define the argument of this associated function). Furthermore, the MEDIAN()

function can be used to calculate the median, and the MODE() function can be used to calculate the mode. At times, MODE() will yield #NA. This simply means that no one gave the same answer more than once and the mode is not a good measurement of the central tendency.

SAMPLE DEVIATION

As noticed in the above plots, the mean may describe the central tendency of a dataset, but many scores will lie above the mean while others lie below it. **To describe the variation of a score about the mean, we calculate the deviation. The deviation is the difference between the given score and the mean.**

$$Deviation = X_i - \bar{X}$$

Deviations will be both positive and negative.

For the entire dataset, it is often useful to consider the variation about the mean in general, rather than simply the deviation of a single data-point away from the mean. The sample standard deviation is a widely used metric of the variation in data about the mean.

To calculate the sample standard deviation, we sum the squares of the individual score's deviation, divide by the number of items in the sample less one, and then take the square root of this quantity. It is common to denote the sample's standard deviation with "s".

$$s = \sqrt{\frac{\sum_{i=1}^{N}(X_i - \bar{X})^2}{N - 1}}$$

Why do we square the terms then take the square root? There are many reasons, but here is a very simple one. If we just added the deviations together, we would always get zero, since some are positive and others are negative, and the average deviation will be zero – every time. However, if we square the deviations, we will always end up with a positive number, and hence the sum of the squares will be some positive number. By taking the square root of the sum of the squares, we are reducing the metric of standard deviation back to the same dimensions as the data itself. Hence, we end up with a useful metric of the variation in the data.

For a worked example, consider the dataset of the distance traveled to school. We would find the following deviations and deviation squared. We use the "^" symbol in Excel to state that the number should be raised to a power, for instance, =5.2^2 means "equals 5.2 raised to the second power", or "5.2 squared".

Metrics	Distance Traveled (miles)	Deviation Formula	Deviation Result	Deviation Squared Formula	Deviation Squared Result
Sum	268.9		0		2357.6
Count	31				
Average	8.7		0		
	3.5	=3.5-8.7	-5.2	=-5.2^2	26.8
	3.2	=3.2-8.7	-5.5	=-5.5^2	30.0
	18.7	=18.7-8.7	10.0	=10.0^2	100.5
	2.1	=2.1-8.7	-6.6	=-6.6^2	43.2
	0.9	=0.9-8.7	-7.8	=-7.8^2	60.4
	14.3	=14.3-8.7	5.6	=5.6^2	31.6
	4.3	=4.3-8.7	-4.4	=-4.4^2	19.1
	8.2	=8.2-8.7	-0.5	=-0.5^2	0.2
	10.8	=10.8-8.7	2.1	=2.1^2	4.5
	22.4	=22.4-8.7	13.7	=13.7^2	188.4
	0.5	=0.5-8.7	-8.2	=-8.2^2	66.8
	1.5	=1.5-8.7	-7.2	=-7.2^2	51.5
	7.3	=7.3-8.7	-1.4	=-1.4^2	1.9
	6.4	=6.4-8.7	-2.3	=-2.3^2	5.2
	24.6	=24.6-8.7	15.9	=15.9^2	253.6
	15.2	=15.2-8.7	6.5	=6.5^2	42.6
	0.5	=0.5-8.7	-8.2	=-8.2^2	66.8
	19.6	=19.6-8.7	10.9	=10.9^2	119.4
	3.6	=3.6-8.7	-5.1	=-5.1^2	25.7
	28.3	=28.3-8.7	19.6	=19.6^2	385.2
	3.1	=3.1-8.7	-5.6	=-5.6^2	31.1
	3.5	=3.5-8.7	-5.2	=-5.2^2	26.8
	2.8	=2.8-8.7	-5.9	=-5.9^2	34.5
	4.2	=4.2-8.7	-4.5	=-4.5^2	20.0
	3.5	=3.5-8.7	-5.2	=-5.2^2	26.8
	3.1	=3.1-8.7	-5.6	=-5.6^2	31.1
	4.3	=4.3-8.7	-4.4	=-4.4^2	19.1
	34.1	=34.1-8.7	25.4	=25.4^2	646.5
	3.4	=3.4-8.7	-5.3	=-5.3^2	27.8
	7	=7-8.7	-1.7	=-1.7^2	2.8
	4	=4-8.7	-4.7	=-4.7^2	21.8

The total distance traveled by these 31 students would be 268.9 miles, (SUM() of all the distances traveled).

The mean distance traveled would be given by

$$\bar{X} = \frac{\sum_{i=1}^{N} X_i}{N} = \frac{SUM(X_I)}{COUNT(X_I)} = \frac{268.9}{31} = 8.7$$

The square of the deviations in distance traveled by these 31 students would be 2411.8 miles, (SUM() of the deviations squared).

The sample's standard deviation in distance traveled would be given by

$$s = \sqrt{\frac{\sum_{i=1}^{N} (X_i - \bar{X})^2}{N-1}} = SQRT\left\{\frac{SUM[(X_I - \bar{X})^2]}{COUNT(X_I) - 1}\right\} = SQRT\left\{\frac{2411.8}{31-1}\right\} = SQRT\{80.4\} \cong 9.0$$

Excel also has a function, STDEV(), that calculates the standard deviation of a sample dataset.

Mean, Median, and Mode from Frequency Distributions

While the above equations and functions define averages and standard deviations, they can only be used directly when raw response data is provided. For instance, the equation and Excel function for averages assumes that each data point represents a single response. However, marketing analysts often receive datasets without raw data. Sometimes, for example, they only receive frequency distributions. To calculate averages and standard deviations from frequency distributions, these equations must be recast using the concept of probability weighted averages.

For instance, consider the responses to a simple survey question. A small sample of nine people were asked: "How much do you agree with the following statement?"

	Strongly Disagree	Disagree	Neither Agree nor Disagree	Agree	Strongly Agree
Irish music is fun					

After collecting data, response frequencies were tabulated and the level of agreement was encoded on a 1 to 5 scale with 1 representing "Strongly Disagree" and 5 representing "Strongly Agree".

	Strongly Disagree	Disagree	Neither Agree nor Disagree	Agree	Strongly Agree
ENCODING	1	2	3	4	5
Irish music is fun	0	0	1	5	3

By reviewing this frequency distribution, we can tell that the mode response is a 4. More people reported "Agree" than any other response. Similarly, by reviewing this frequency distribution, we would determine that the median response is 4, as 4 people gave a response of 4 or less and 4 people gave a response of 4 or more. Hence, 4 looks like a good measurement of the central tendency of this distribution. But how would we determine the mean response?

If we took the average encoding, we would get a 3. However, 3 doesn't represent the average response, it only represents the middle of the possible responses. (In the same way, a 50 on a 100 point test rarely represents the average test score). Likewise, if we took the average response frequency, we would get 1.8. However, 1.8 doesn't represent the average response, it only represents how often on average a response is given.In this case, several of the possible responses were never given.

To find the average response, we could identify how many times each response is scored. In the sample dataset, we had 0 responses of score 1, 0 responses of score 2, 1 response of score 3, 5 responses of score 4, and 3 responses of score 5. Hence, the scores include 1 three, 5 fours, and 3 fives.

Scores: 3, 4, 4, 4, 4, 4, 5, 5, 5,

Using the formulas for average and standard deviations on these scores, we find the average to be 4.2 with a standard deviation of 0.7.

While this brute force approach can be used, there are simpler solutions. Effectively, we are weighting the response scores with the response frequencies. That is, we are finding the frequency weighted average score. Thus, the weighted average could be given by

$$\bar{X} = \frac{(1 \cdot 0) + (2 \cdot 0) + (3 \cdot 1) + (4 \cdot 5) + (5 \cdot 3)}{1 + 5 + 3}$$

$$\bar{X} = 4.2$$

We can generalize this approach for calculating the frequency weighted average. If f_i (read as "f sub I") represents the frequency of response i, and X_i is the score for response i, the frequency weighted average response is given by the following expression

$$\bar{X} = \frac{\sum_i X_i \cdot f_i}{\sum_i f_i}$$

Using a similar notation, we would find the sample's standard deviation as

$$s = \sqrt{\frac{\sum_i (X_i - \bar{X})^2 \cdot f_i}{\sum_i f_i - 1}}$$

Using this formula, we would once again find the standard deviation of the weighted average response to be 0.7.

We can use a very similar formula for calculating the probability weighted average response. In this case, we would use

$$\bar{X} = \sum_i X_i \cdot P_i$$

where P_i is the probability of obtaining a specific response.

We cannot calculate the standard deviation with probabilities alone. We must know the sample size to calculate the standard deviation from probabilities. If we know the sample size, the standard deviation can be calculated as

$$s = \sqrt{\sum_i (X_i - \bar{X})^2 \cdot P_i \cdot \frac{N}{N-1}} \cong \sqrt{\sum_i (X_i - \bar{X})^2 \cdot P_i} \; for\ large\ N$$

Marketing Metrics: Profit Equation of the Firm

Marketing has a strategic responsibility for the profitability of the firm. The core decisions in marketing determine price, promotions, placement, and product strategy. Therefore, marketing and marketers influence demand (also known as quantity sold), production costs, and overhead costs. Price, demand, variable costs, and fixed cost are all determinants of profits – and marketing directly or indirectly manages each of these variables. Therefore, the profitability of the firm is the strategic responsibility of marketing.

Because marketers are responsible for ensuring the profitability of the firm, it is imperative that marketers are prepared to understand the profit impact of their decisions on the firm and forecast the firm's profitability.

Revenue

Revenue is the amount of money the firm makes over a given period. On a profit and loss statement, it appears at the top and is often referred to as "top line revenue".

If a shop sells French fries for $1.50, and sells 2,234 orders of French fries in a given month, then the revenue earned from French fries is $1.50 per order X 2,234 orders equal to $3,351.

In general, we would state

Revenue = Price X Quantity Sold

Or, using R for Revenue, P for Price, and Q for Quantity Sold, we could write

R = P • Q

Costs

Costs are the amount of money the firm spends over a given period, presumably to earn the revenue generated on the top line of the profit and loss statement. Costs appear in the middle of a profit and loss statement.

In general, firms incur two types of costs: fixed and variable.

- **Fixed costs are costs which are constant regardless of the number of units sold. Typical fixed costs items include property, plant, equipment, management salaries, etc.**
- **Variable costs are costs which increase with the number of units sold. Typical variable costs items include input costs, such as raw materials and sometimes labor.**

If the above French fry maker had to pay $2000 per month for rent, heat, and electricity, and each order of French fries used $0.25 in potatoes and oil, we would state that the fixed costs were $2000 per month and the variable costs were $.25 per order.

The cost of goods sold (COGS) represents the total variable costs in a given period.

Our French fries shop, selling 2,234 orders of French fries in a given month, would find the COGS at $.25 per order times 2,234 orders equal to $558.50.

In general, we would state

Cost of Goods Sold = Variable Costs X Quantity Sold

Or, using COGS for Cost of Goods Sold, VC for variable costs, and Q for Quantity Sold, we could write

COGS = VC • Q

Moreover, we would state that the total costs in general are

Total Costs = Cost of Goods Sold + Fixed Costs

Or, using FC for Fixed Costs and TC for Total Costs, we would write

TC = COGS + FC

or

TC = VC • Q + FC

Our French fries shop, selling 2,234 orders in a given month with variable costs of $.25 per order and fixed costs of $2000 per month, would face total monthly costs of $2,558.50.

PROFIT

Profit is the amount of money a firm earns. It is the difference between the revenue earned from meeting customer needs and the costs the firm incurs delivering products and services to meet their needs. Hence, profit is simply revenue less costs.

Profit = Revenue – Costs

Profit is also referred to as Gross Margin. Profits, or gross margins, appear at the bottom line of the firm's profit and loss statement, and are often referred to as "bottom line profit".

Using the above components for revenue and costs, we could calculate the profit equation of the firm using any of the forms below.

Profit = R – TC

Profit = R – COGS – FC

Profit = P • Q – VC • Q – FC

And, after rearranging the last equation, we find the profit equation of the firm.

Profit = Q • (P– VC) – FC

While the last form of the profit equation will be heavily used in this course, students should be prepared to use any of the above forms of the firm's profit equation when analyzing a strategic marketing question.

Our French fries shop selling 2,234 orders of French fries at $1.50 in a given month with variable costs of $.25 per order and fixed costs of $2000 per month would enjoy monthly profits of $793. Profits this low might not be sufficient to satisfy the shop keeper.

EXERCISES

KALE

A survey was conducted asking for the level of agreement among a sample. The following data was collected and tabulated.

Response Items	Strongly Disagree	Disagree	Neither Agree nor Disagree	Agree	Strongly Agree
ENCODING	1	2	3	4	5
I like Kale.	2	7	5	4	3

1. How many people were surveyed?
2. Find the percentage of responses in each category.
3. Create a column chart of the probability of a given response. Label the horizontal axis with the response items. Title the graph with "Level of Agreement with 'I like Kale'". Title both the horizontal and vertical axis appropriately.
4. By examining the graph, how would you describe the skewness of the distribution: Strongly negatively skewed, somewhat negatively skewed, relatively unskewed, somewhat positively skewed, strongly positively skewed?
5. By examining the graph, how would you describe the kurtosis of the distribution: strongly leptokurtic, strongly platykurtic, or strongly mesokurtic, between leptokurtic and mesokurtic, or between platykurtic and mesokurtic.
6. What is the median response?
7. What is the mode response?
8. What is the weighted average response?
9. What is the standard deviation of the weighted average response?

MEN'S SHOES

A sample of males was surveyed regarding their shoe sizes and data was collected.

1. Create a column chart of the probability of a given response. Title the graph and all axes appropriately.
2. By examining the graph, how would you describe the skewness of the distribution: Strongly negatively skewed, somewhat negatively skewed, relatively unskewed, somewhat positively skewed, strongly positively skewed?
3. By examining the graph, how would you describe the kurtosis of the distribution: strongly leptokurtic, strongly platykurtic, or strongly mesokurtic, between leptokurtic and mesokurtic, or between platykurtic and mesokurtic.
4. What is the median men's shoe size?
5. What is the mode men's shoe size?
6. What is the weighted average men's shoe size?
7. What is the standard deviation of the weighted average men's shoe size?
8. The market analyst decided they needed to recast this shoe size distribution for ordering socks in small, medium, and large. Given these three intervals:
    a. Which sizes would fit in each interval?
    b. What is the percentage of respondents that fit in each interval?
    c. Create a percentage plot of the men's shoe sizes with three intervals.

Male Shoe Size	Response Frequency
Unknown	6
6	17
7	2
8	51
9	154
10	172
11	94
12	51
13	21
14	5

HELLO KITTY

A survey was conducted asking for the level of agreement among a sample. The following response probabilities were collected among 198 high school teenagers.

Response Items	Strongly Disagree	Disagree	Neither Agree nor Disagree	Agree	Strongly Agree
ENCODING	1	2	3	4	5
I like Hello Kitty.	5%	14%	29%	33%	19%

1. Create a column chart of the probability of a given response. Label the horizontal axis with the response items. Title the graph and both axes appropriately.
2. By examining the graph, how would you describe the skewness of the distribution: Strongly negatively skewed, somewhat negatively skewed, relatively unskewed, somewhat positively skewed, strongly positively skewed?
3. By examining the graph, how would you describe the kurtosis of the distribution: strongly leptokurtic, strongly platykurtic, or strongly mesokurtic, between leptokurtic and mesokurtic, or between platykurtic and mesokurtic?
4. What is the median response?
5. What is the mode response?
6. What is the weighted average response?
7. What is the standard deviation of the weighted average response?

All Souls Gathered

The following data regarding the number of souls gathered at Hyde Park Union Church for a number of Sundays has been collected.

1. How many Sundays are in the dataset?
2. Calculate and label each of the following figures.
    a. Maximum attendance.
    b. Minimum attendance.
    c. Range of the attendance.
    d. Average attendance.
    e. Standard deviation in attendance.
    f. Mode attendance.
    g. Median attendance.
3. If we wanted to use 5 intervals to sort all the data:
    a. What is the minimum size of each interval?
    b. What is a natural interval size to use that is larger than the minimum acceptable interval size?
    c. Where should the lowest interval start?
4. Find the percentage of Sundays with attendance in each of the 5 intervals and create a percentage plot.
    a. Create columns describing the upper limits and lower limits of each interval.
    b. Use the Histogram dialogue in Excel to create a frequency distribution of the attendance.
    c. Calculate the percentage of Sundays in each interval.
    d. Create a column containing the horizontal axis labels for intervals.
    e. Create a column chart of the percentage of weeks in each interval. Title each axis and graph.
5. By examining the graph, how would you describe the skewness of the distribution: Strongly negatively skewed, somewhat negatively skewed, relatively unskewed, somewhat positively skewed, strongly positively skewed?
6. By examining the graph, how would you describe the kurtosis of the distribution: strongly leptokurtic, strongly platykurtic, or strongly mesokurtic, between leptokurtic and mesokurtic, or between platykurtic and mesokurtic .

All Souls Gathered								
68	89	105	102	75	95	114	95	88
89	97	79	102	93	76	100	152	105
66	93	102	81	135	78	92	110	91
80	83	108	72	110	125	103	105	83
61	87	90	80	106	80	82	99	144
66	90	110	59	97	102	107	111	106
87	106	110	85	82	83	189	115	120
72	97	115	96	107	114	92	112	118
55	90	129	92	180	107	73	130	73
108	194	97	73	106	98	61	146	115
69	111	47	54	97	108	70	67	113
101	75	102	97	97	103	71	128	126
115	89	111	121	94	115	65	100	99
92	82	94	111	112	90	75	115	127
100	98	84	113	104	127	93	115	139
93	83	99	97	105	128	71	95	175
107	71	82	98	81	97	100	100	130
104	80	102	102	110	63	111	120	88
95	71	117	117	105	104	90	115	117
97	64	99	146	73	118	115	127	121
108	80	106	112	75	85	101	96	130
93	100	105	104	76	99	102	94	80
135	67	110	92	75	101	122	95	54
93	69	175	152	87	114	98	120	78
99	99	81	97	72	112	95	113	105
124	104	100	95	58	134	95	101	115
75	88	90	84	78	91	97	105	110
88	57	118	98	100	67	105	95	120

ComEd Electric Bills

Monthly winter ComEd residential electric bills from various households in Chicago have been collected.

1. How many bills are in the dataset?
2. Calculate and label each of the following figures:
    a. Maximum bill.
    b. Minimum bill.
    c. Range of in the bill amount.
    d. Average bill.
    e. Standard deviation in the bill amount.
        i. Deviation for each point.
        ii. Deviation squared for each point.
        iii. Sum of deviations squared.
        iv. Sum of deviations.
        v. The standard deviation using the formula for the square root of the sum of the deviations squared divided by the number of items less one.
    f. Mode bill.
    g. Median bill.
3. If we wanted to use 7 intervals to sort all the data:
    a. What is the minimum size of each interval?
    b. What is a natural interval size to use that is larger than the minimum acceptable interval size?
    c. Where should the lowest interval start?
4. Find the percentage of bills within each of the 7 intervals and create a percentage plot.
    a. Create columns describing the upper limits and lower limits of each interval.
    b. Use the Histogram dialogue in Excel to create a frequency distribution of the bills.
    c. Calculate the percentage of bills in each interval.
    d. Create a column containing the horizontal axis labels for interval.
    e. Create a column chart of the percentage of bills in each interval. Title each axis and graph.
5. By examining the graph, how would you describe the skewness of the distribution: Strongly negatively skewed, somewhat negatively skewed, relatively unskewed, somewhat positively skewed, strongly positively skewed?
6. By examining the graph, how would you describe the kurtosis of the distribution: strongly leptokurtic, strongly platykurtic, or strongly mesokurtic, between leptokurtic and mesokurtic, or between platykurtic and mesokurtic .

Electric Bills								
36	23	36	30	36	30	38	45	58
38	32	52	49	45	36	59	52	29
49	44	38	55	54	35	53	49	35
34	36	55	41	78	24	43	48	38
24	41	39	36	40	73	52	36	38
21	44	31	24	40	28	44	39	68
27	35	64	31	33	44	46	41	42
51	51	45	47	25	48	75	40	61
29	43	60	53	48	64	46	39	60
25	43	52	50	54	53	27	41	33
47	96	35	23	39	48	19	44	52
36	55	20	20	52	56	22	26	43
38	41	33	49	39	45	35	66	38
47	38	38	37	38	36	35	30	38
47	35	42	49	54	48	35	45	99
50	53	27	49	48	74	52	35	63
46	48	45	31	59	49	23	32	78
58	30	46	50	33	37	48	43	80
49	45	35	43	51	31	62	58	39
32	39	59	54	49	49	53	54	68

Firm Performance

Sales performance data over 26 weeks for a firm were collected. Each unit was priced at $14.99.

1. Calculate:
    a. The revenue earned each week.
    b. The maximum revenue in a given week.
    c. The minimum revenue in a given week.
    d. Mean weekly revenue.
    e. Median weekly revenue.
    f. Standard deviation in weekly revenue.
2. Make a probability distribution plot of the revenue per week using an interval of $50,000, with the smallest group having a lower limit of $100,000.
3. Describe the skewness of the dataset.
4. Describe the kurtosis of the dataset.
5. Are there any outliers in the dataset?
6. Which is more representative of the central tendency, mean or median?
7. If the cost to produce each unit sold was $2.75 and the weekly fixed costs are $150,000, calculate:
    a. The cost of goods sold (COGS) each week.
    b. The total cost for each week.
    c. The profitability for each week.
8. Profit analysis
    a. Did the company make profits every week?
    b. Using a "Line Plot", plot the weekly profit versus the week to demonstrate how profits change over the time period.
    c. Over the time period in question, were profits increasing or decreasing in general?
    d. Make a probability distribution plot of the firm's profits using 5 intervals.
9. Calculate
    a. Total revenue over the 26 weeks.
    b. Total costs over 26 weeks.
    c. Total profitability over the 26 weeks.
10. Imagine you were reporting this data to an executive.
    a. What might you suggest needs exploring to explain the outliers in the dataset?
    b. What might you suggest is driving the reduction in profits, too few unit sales or too many price reductions?

Week	Unit Sales	Week	Unit Sales
1	10,662	14	14,180
2	15,119	15	15,398
3	16,751	16	19,882
4	13,493	17	15,951
5	9,573	18	14,732
6	14,284	19	28,056
7	15,244	20	15,375
8	22,138	21	16,843
9	34,342	22	12,795
10	18,446	23	13,131
11	11,544	24	11,004
12	11,940	25	14,206
13	10,162	26	10,440

CHAPTER 3: CROSS TABS AND SAMPLING

When dealing with categorical data, the distribution of scores is often examined in a cross tabulation table and depicted with stacked bar or stacked column charts. We have already seen a number of individual cross-tab tables in prior chapters. However, we haven't yet compared them across samples. Different samples may have different distributions. **A cross tab analysis that includes stacked bar and column charts is very useful for comparing distributions across samples.**

In this chapter, students will learn:

- How to create a simple cross tab table given raw response data using the Excel functions of IF(), LOOKUP() and COUNTIF().
- How to create a cross tab table given raw response data using the Excel sort tool and functions of LOOKUP() and COUNTIF().
- How to create a stacked bar and column chart of the cross tab data in Excel.
- How to compare cross-tab data across samples.
- Aspects of sampling design, including sample size and sample representation.
- The difference between margins and mark-ups.

CROSS TABS WITH SINGLE SAMPLES AND ONLY 2 CATEGORIES

Consider a web marketer testing a page layout. They collect the following series of events.

Purchase or No Purchase			
Purchase	Purchase	No Purchase	No Purchase
No Purchase	No Purchase	Purchase	Purchase
No Purchase	No Purchase	No Purchase	No Purchase
No Purchase	No Purchase	No Purchase	No Purchase
No Purchase	No Purchase	No Purchase	No Purchase
No Purchase	No Purchase	No Purchase	

The research would like to know the percentage of times a person purchases.

While it is possible to hand count the number of times a person did or did not purchase an item, a less tedious and error prone approach would be to use an IF() function in Excel. The IF() function in Excel has three arguments: the expression to be tested for validity, the value to provide when the expression is true, and the value to be provided when the expression is false.

If the researcher assigns the value of 1 when a customer purchases, and a value of 0 when a customer doesn't purchase, then the percentage of purchasing can easily be found by adding up all the ones over the dataset, and dividing by a count of the elements in the dataset.

The following steps can be used with the IF(), SUM(), and COUNT() functions to create a cross tab of this data. Assume the data lies in cells A1 through D6.

- In cell A8, type "=if(A1="Purchase",1,0)". This function will compare the value in cell A1 with the expression "Purchase" and determine if they are equal or not. If they are equal, then the expression is valid and the function will return the "Value If True" of 1. If they are not equal, then the expression is invalid and the function will return the "Value If False" of 0.
- Copy and paste this expression throughout the remaining cells required to encode all the data points.
- The result should appear similar to that below:

	A	B	C	D
1	Purchase	Purchase	No Purchase	No Purchase
2	No Purchase	No Purchase	Purchase	Purchase
3	No Purchase	No Purchase	No Purchase	No Purchase
4	No Purchase	No Purchase	No Purchase	No Purchase
5	No Purchase	No Purchase	No Purchase	No Purchase
6	No Purchase	No Purchase	No Purchase	
7				
8	1	1	0	0
9	0	0	1	1
10	0	0	0	0
11	0	0	0	0
12	0	0	0	0
13	0	0	0	

At this point, the sum of all the codes will equal the purchase frequency, a count of all the codes will equal the total number of observations made, and the count less the sum will yield the number of losses.

- Title cells A 16, 17, and 18 as "Purchase", "No Purchase", and "Total "respectively. Title cells B15 and C15 "Frequency" and "Probability" respectively.
- In cell B16, type "=sum(A8:D13)", the sum of the range of cells with the encoded values.
- In cell B18, type "=count(A8:D13)", the count of the range of cells with the encoded values.
- In cell B17, type "=B18-B16", the difference of the count and the sum.
- You now have the frequencies. Calculate the probabilities appropriately.
- The results should appear similar to that shown below.

	A	B	C	D
14				
15		Frequency	Probability	
16	Purchases	4	17%	
17	No Purchases	19	83%	
18	Total	23	100%	
19				

Cross Tabs with Single Samples and Multiple Categories

Consider the survey responses of a sample of 44 students in front of DePaul. They were asked "On average, how often would you buy a hotdog if it were available in front of the classrooms at a reasonable price?" The possible responses were "Never, once a year, once a month, once a week, two or more times a week". In tabulating the survey responses, the researcher gathered a number of responses.

Respondent	Response	Respondent	Response
1	Once a Month	23	Once a Week
2	Once a Month	24	Once a Month
3	Once a Year	25	Once a Month
4	Two or More Times a Week	26	Never
5	Once a Month	27	Once a Week
6	Once a Month	28	Two or More Times Week
7	Once a Month	29	Once a Month
8	Once a Year	30	Once a Week
9	Once a Month	31	Once a Week
10	Once a Week	32	Never
11	Never	33	Once a Month
12	Once a Year	34	Two or More Times a Week
13	Never	35	Once a Month
14	Once a Week	36	Once a Week
15	Once a Month	37	Once a Month
16	Once a Week	38	Once a Week
17	Once a Month	39	Once a Year
18	Never	40	Once a Month
19	Once a Week	41	Once a Year
20	Once a Week	42	Once a Month
21	Once a Year	43	Once a Month
22	Two or More Times a Week	44	Never

The researcher can count how often a given response is made to measure the response frequency, and, after dividing by the total number of responses, the researcher will have a measure of the response probability gathered from that sample.

Manually counting each response to a large survey, however, is tedious and error prone. Once again, Excel provides a number of techniques for converting raw data into meaningful metrics. Two functions of particular use in creating cross-tabs are LOOKUP() and COUNTIF().

The LOOKUP() function allows a researcher to look up a value in a table and determine the score that should be associated with that value. Hence, it must be used in conjunction with an encoding table. The LOOKUP() function has arguments of Value, Lookup Vector, and Result Vector. Value refers to the value that must be encoded. The Lookup Vector is a column range containing the possible values to be encoded. Importantly, the Lookup Vector array is put in alpha-numeric order. Result Vector contains the code to be associated with each value.

For an example of LOOKUP() in Excel, consider an analyst who collects the previous hot dog eating frequency data and places the data in columns A and B in Excel. Column A contains the respondent I.D. and column B contains the text response. In Column D the analyst types in the list of possible responses. Right next to the list of possible responses in Column E, the analyst inputs a list of the relevant codes, 5 for two or more a week, 4 for once a week, 3 for once a month, 2 for once a year, and 1 for never. The result of this preparatory step might look like this.

	A	B	C	D	E
1	Respondent id	Response		Response Value	Code
2	1	Once a Month		Two or More Times a Week	5
3	2	Once a Month		Once a Week	4
4	3	Once a Year		Once a Month	3
5	4	Two or More Times a Week		Once a Year	2
6	5	Once a Month		Never	1
7	6	Once a Month			
8	7	Once a Month			
9	8	Once a Year			
10	9	Once a Month			
11	10	Once a Week			
12	11	Never			
13	12	Once a Year			

As stated, the LOOKUP() function requires the Values Vector to be in alpha-numeric order. To sort the possible response values and codes in alpha-numeric order by response values:

- Select the data that will be in the lookup vector and lookup value only, in this case cells D2 through E6.
- Go to the "Data" tab, hit the "Sort" icon, and a sorting dialogue should appear.
  - Ensure that the "My Data Has Headers" box is selected if your data has headers.
  - Select the "Column Sort By" to equal "Response". Leave the "Sort On" as "Values". Select the sort order as "A to Z" for ascending alpha-numeric order.
  - Hit ok.
- The results should appear as follows:

	D	E
1	Response Value	Code
2	Never	1
3	Once a Month	3
4	Once a Week	4
5	Once a Year	2
6	Two or More Times a Week	5
7		

Now the analyst is ready to use the LOOKUP() function.

The analyst chooses to title column C with "R-Code" for response code. Then, in cell C2, the analyst types "=LOOKUP(B2,D$2:D$6,E$2:E$6)". Cell B2 contains the value to be examined in the lookup table. Cells D2 through D6 contain the lookup values for encoding. Cells E2 through E6 contain the result vector of the encoding. The dollar signs are used in front of the row numbers describing the lookup vector and results vector so that this formula can be copied from cell to cell. After copying and pasting this formula for all the cells to be encoded, the researcher generates the following data.

	A	B	C	D	E
1	Respondent id	Response	R-Code	Response Value	Code
2	1	Once a Month	3	Never	1
3	2	Once a Month	3	Once a Month	3
4	3	Once a Year	2	Once a Week	4
5	4	Two or More Times a Week	5	Once a Year	2
6	5	Once a Month	3	Two or More Times a Week	5
7	6	Once a Month	3		
8	7	Once a Month	3		
9	8	Once a Year	2		
10	9	Once a Month	3		
11	10	Once a Week	4		
12	11	Never	1		
13	12	Once a Year	2		

The dataset has now been encoded, yet it still doesn't reveal the probability of a specific response being elicited from a person. The next step in creating a cross-tab requires the use of the COUNTIF() function.

The COUNTIF() function is used to count the number of times a specific value appears. It has two arguments, Range and Value. The Range argument will define the array of data to be examined for matching values. The Value argument is a single value that is to be matched. Each time a cell in the Range matches the desired Value, the count increases by one.

For an example of COUNTIF() function in Excel, let us build on our hot dog example.

- Title column F "Response Frequency". Column F will contain our response frequencies.
- In cell F2, type "=COUNTIF(C$2:C$45,E2)" where cells C2 through C45 contain the encoded data, and cell E2 contains the value to be counted. Once again, the dollar sign has been used in front of the row numbers of the array containing the encoded values to be counted.
- Copy and paste this formula down to span the set of possible codes.
- The result should look as follows.

	B	C	D	E	F
1	Response	R-Code	Response Value	Code	Response Frequency
2	Once a Month	3	Never	1	6
3	Once a Month	3	Once a Month	3	17
4	Once a Year	2	Once a Week	4	11
5	Two or More Times a Week	5	Once a Year	2	6
6	Once a Month	3	Two or More Times a Week	5	4
7	Once a Month	3			
8	Once a Month	3			
9	Once a Year	2			
10	Once a Month	3			
11	Once a Week	4			
12	Never	1			
13	Once a Year	2			

At this point, a frequency distribution has been created, but it isn't in a rational format for presenting as a cross tab. Specifically, the categories are not listed in a rational format. They need to be sorted back into the original order. Unfortunately, sorting them *in situ* on this spreadsheet will muck up the calculations used to derive the current results. Specifically, the LOOKUP() results will be miscalculated if the data is altered. As such, we shall copy and paste these results into a new spreadsheet as values.

- Select cells D1 through F6 and copy.
- On a new spreadsheet, paste these cells as values. Pasting cells as values requires selecting the "1,2,3 VALUES" button from the "Paste Drop Down" menu.
- Now, sort the data by Code value from smallest to largest.
- You should end up with the following result.

	A	B	C	D
1	Response	Code	Response Frequency	
2	Never	1	6	
3	Once a Year	2	6	
4	Once a Month	3	17	
5	Once a Week	4	11	
6	Two or More Times a Week	5	4	
7				

After calculating percentages for each response, the following cross tab table is created.

Response	Frequency	Percentage
Never	6	14%
Once a Year	6	14%
Once a Month	17	39%
Once a Week	11	25%
Two or More Times a Week	4	9%
Sum	44	100%

Next, a stacked bar chart can be used to represent the frequency distribution. After inserting a blank 100% Stacked Bar Chart, the data must be selected. In a stacked bar or column chart, each response item will be an independent data series.

- Hit "Select Data" for your blank Stacked Bar Chart.
- Add a new series.
- For Series name, select the "Never" cell, A2.
- For the series value, select the response frequency cell, C2.
- Repeat for the remaining possible responses.
- For the Horizontal Axis Label, choose the cell title "Response".
- After selecting a proper format, the result should be similar to that shown below.

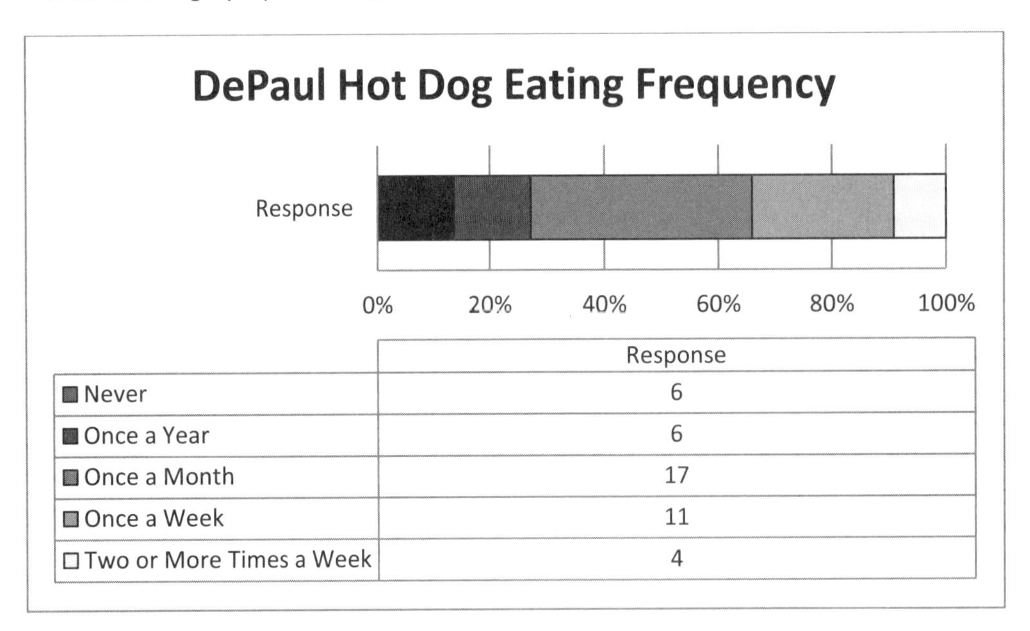

Cross Tabs with Multiple Samples and Multiple Categories

Creating a cross tabs and stacked bar chart with multiple samples requires similar steps to that for a single sample. The key difference lies in selecting the data to be plotted.

Suppose the hot dog eating frequency was also measured among business people in front of the Chicago Federal Reserve building and the following response frequencies were collected, tabulated, and displayed next to the DePaul data.

	A	B	C	D
1	Response	DePaul	Federal Reserve	
2	Never	6	7	
3	Once a Year	6	8	
4	Once a Month	17	23	
5	Once a Week	11	6	
6	Two or More Times a Week	4	2	
7				

To compare this data, the analyst again inserts a blank 100% Stacked Bar Chart.

- Hit "Select Data" for your blank Stacked Bar Chart.
- Add a new series.
- For Series name, select the "Never" cell, A2.
- For the series value, select the response frequencies cells, those for both the DePaul and Federal Reserve samples, cells B2:C2.
- Repeat for the remaining possible responses.
- For the Horizontal Axis Label, choose the cells titled "DePaul" and "Federal Reserve", cells B1:C1. If multiple samples are used, the Horizontal Axis Labels can be used to identify specific samples.
- After selecting a proper format, the result should be similar to that shown below.

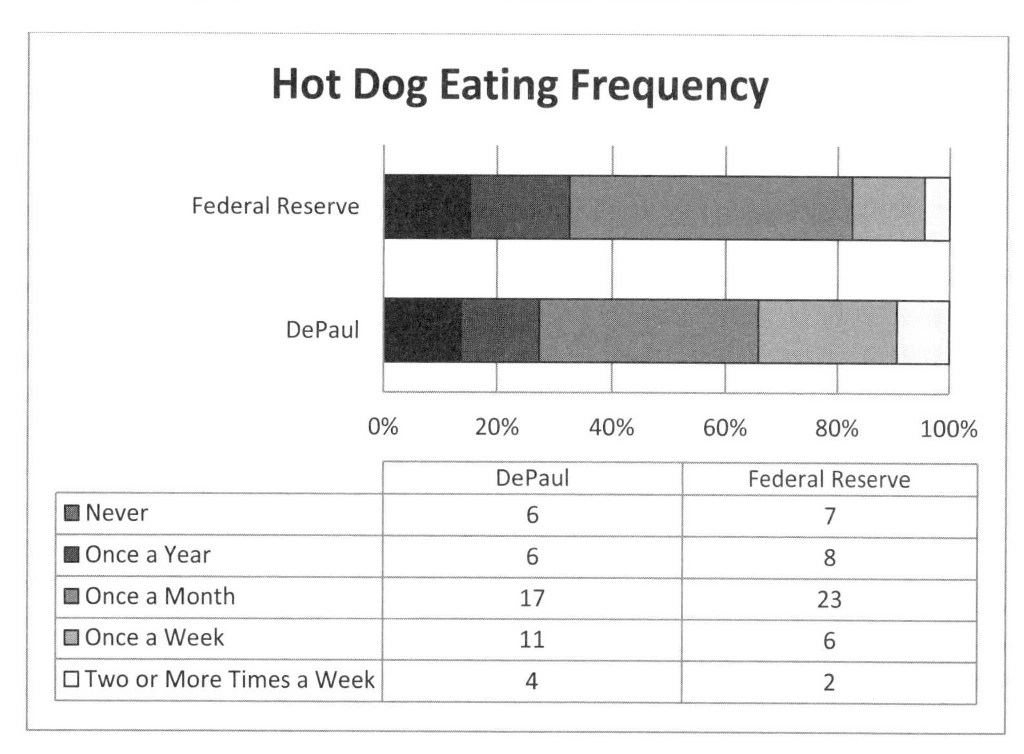

	DePaul	Federal Reserve
Never	6	7
Once a Year	6	8
Once a Month	17	23
Once a Week	11	6
Two or More Times a Week	4	2

From this analysis, it is clear that people near the Federal Reserve eat fewer hot dogs than those near DePaul.

To calculate the category interaction frequencies of these two samples, we must interpret the meaning of the responses. On an annual basis, the following interpretations are used.

Response	Annual Category Interaction
Never	0
Once a Year	1
Once a Month	12
Once a Week	52
Two or More Times a Week	104

Calculating the category interaction for both the DePaul sample and the Federal Reserve sample, we find that the average DePaul student reports eating 27 hot dogs a year, while an average business person in front of the Federal Reserve only eats 17 hot dogs a year.

Sampling

While it would be nice to measure the entire population's response to a query, such an approach is unrealistic. For example, the US spent over $14 billion just counting the US population in 2010 census. No researcher finds this approach cost effective.

Instead, research must rely on samples. While market researchers accept that samples are not the total population, we can take measures to improve our sampling.

At a high level, there are two key dimensions to concern oneself with when sampling: sample size and sample representation.

Size

Sample size refers to the actual number of sample units within the sample. In general, a better representation of the population is obtained from a larger sample than a smaller sample.

Unfortunately, the improvement increases with the square root of the sample size. That is, suppose 1000 people were surveyed and asked "Should the US open its boarders to Canada?" If 51% said yes and 49% said no, that would imply that the majority agree with this proposal. However, the margin of error with a sample of this size is 3%. That is, we cannot say with certainty that the population's true opinion is 51/49. It could be as low as 48% in favor/52% against, or as high as 54% in favor / 46% against. To improve the accuracy of our measurement by a factor of two (down to a margin of error of 1.5%), we have to increase the sample size by a factor of 4 (up to 4000). Worse, to reduce the margin of error to within a single percentage point of 0.75%, we would have to increase the sample size another factor of 4 to 16,000 individuals. We quickly begin to realize that increasing the sample size is not always a reasonable course of action.

Increasing the sample size is almost always possible, but the costs of doing so should be weighed against the value of improving the accuracy of the research effort.

In many cases, a small but representative sample provides sufficient insight for management decision making. In almost all circumstances, any sample of actual customers is likely to provide more definitive insights than no sample at all.

Representation

Sample representation refers to the ability of the members within the sample to represent the population as a whole. In general, statistical results from a sample provide a better representation of the population when the make-up of the sample more closely matches that of the population.

When the sample systematically misrepresents the population, we have a biased sample.

Incidental sampling, also known as convenience sampling, can be representative in some cases and biased in others. Incidental samples are often used because they are easy to collect.

Other forms of sampling include simple random, stratified, and snowball. Consult a market research text for further information.

Examples of sampling include that are sometimes appropriate, and other times inappropriate, are:

- The students passing by the front of campus.
- The contacts within a marketing database.
- The customers at a particular store location.
- The phone numbers from a phone book.
- The members of a research panel who mailed in a survey.

Researchers can take measures to reduce biases, as students will discover when they take an advanced course in market research.

Mean of Sample Means

If a measurement is done repeatedly, the researcher may find slightly different results each time. We have already seen this in the difference between the hot-dog eating frequencies in front of DePaul and that in front of the Federal Reserve. These differences in sample distributions and means are a natural result of sampling the population as opposed to measuring the population as a whole.

With different sample measurements, it is useful to calculate the mean of sample means. The mean of sample means is the weighted average of the means across the different samples, where the weights are the size of the individual samples.

For instance, suppose the following sample sizes and means were uncovered:

Sample	Sample Size	Sample Mean
1	95	42.5
2	153	43.7
3	78	41.8
4	133	43.3

The mean of means would be calculated as

$$\bar{X} = \frac{(N_1 \cdot X_1) + (N_2 \cdot X_2) + (N_3 \cdot X_3) + (N_4 \cdot X_4)}{N_1 + N_2 + N_3 + N_4}$$

$$\bar{X} = \frac{(95 \cdot 42.5) + (153 \cdot 43.7) + (78 \cdot 41.8) + (113 \cdot 43.3)}{95 + 153 + 78 + 133}$$

$$\bar{X} = 43.0$$

This example, however, only begs the question: What is the actual population mean? We can measure the mean of a sample from the population, and then repeat the measurement on a different sample from the same population, and we will likely get a slightly different mean. As such, we must acknowledge that any mean derived from a sample will imperfectly describe the actual population. As such, we state that the sample has a standard error of the mean, a range about the sample mean in which we anticipate the true mean to lie, but are unsure of exactly where the mean lies within that expected error range.

Hence, in our hypothetical survey of 1000 Americans regarding the opening of our border with Canada, we might state that the average is 51 percent plus or minus 3 percent (51% ± 3%).

Marketing Metrics: Margins and Markups

When pricing a product, manufacturers will be concerned with margins and retailers will be concerned with markups.

The contribution margin of a product is the difference between the price and the variable cost of production as a percentage of the price a product. Hence, the contribution margin (CM) describes the portion of every dollar earned that contributes to profits.

$$CM = \frac{P - VC}{P}$$

- If a product has a 50% contribution margin, 50% of every dollar earned in revenue goes to profits, and 50% goes towards variable costs.
- If a product has a 30% contribution margin, 30% of every dollar earned in revenue goes to profits, and 70% goes towards variable costs.

With this definition of the contribution margin, we can rewrite the profit equation of the firm.

Profit = Q • (P– VC) – FC (profit equation of the firm)

Insert VC = P – P • CM to find

Profit = Q • P • CM – FC

As can be seen from the last equation, profits increase as contribution margins increase.

The markup of a product is the difference between the price and the variable cost to stock as a percentage of the variable costs. Hence, the markup, or how much a product is "marked up" for resale.

$$\boldsymbol{Markup = \frac{P - VC}{VC}}$$

And price can be found from variable costs.

$$P = VC \cdot (1 + Markup)$$

Markups are commonly used by retailers for pricing the tens of thousands of items they stock.

EXERCISES

MORNING OR EVENING SELLING IS BEST?

A salesperson is testing two different approaches to engaging a customer. Prospects in group A were contacted before 10 am in the morning. Prospects in group B were contacted after 3 pm in the afternoon. The success in winning or losing a sale was tracked as "Won" or "Lost".

Use the IF(), SUM(), and COUNT() functions in Excel to complete this exercise.

1. What kind of data was collected: Nominal, Ordinal, Interval, or Ratio?
2. How many win's appear in Group A? In Group B?
3. How large is Group A? Group B?
4. How many losses are in Group A? Group B?
5. What is the win ratio for group A? Group B? (Win Ratios are calculated by finding the percentages of wins in relation to all possible events).
6. Create a 100% stacked bar chart that shows the win and loss frequencies for each group on a single graph.
7. What is the overall win ratio across both groups?
8. Based on this dataset alone, would you suggest that the salesperson concentrate their outbound calling effort in the morning or evening? How confident are you in your answer?

Group A				Group B			
Lost	Lost	Lost	Lost	Won	Lost	Lost	Lost
Lost	Lost	Lost	Won	Lost	Lost	Lost	Lost
Won	Lost	Lost	Won	Lost	Lost	Lost	Lost
Lost	Lost	Lost	Lost	Lost	Lost	Lost	Lost
Lost	Lost	Lost	Lost	Lost	Lost	Lost	Lost
Lost	Won	Lost	Lost	Lost	Lost	Lost	Lost
Lost	Lost	Lost	Lost	Lost	Lost	Lost	Lost
Lost	Won	Lost	Lost	Lost	Lost	Lost	Won
Lost	Won	Lost	Lost	Lost	Lost	Lost	Lost
Lost	Lost	Lost	Lost	Lost	Lost	Won	Lost
Won	Lost	Won	Lost	Lost	Lost	Lost	Lost
Lost	Lost	Lost	Won	Won	Lost	Lost	Lost
Lost	Lost	Won	Lost	Lost	Lost	Lost	Lost
Lost	Lost	Lost	Lost	Lost	Lost	Lost	Lost
Lost	Lost	Lost	Lost	Lost	Lost	Lost	Lost
Lost	Won	Lost	Lost	Lost	Lost	Lost	Lost
Won	Lost	Lost	Won	Lost	Lost	Lost	Lost
Lost	Won	Lost	Lost	Lost	Won	Lost	Lost
Lost	Won	Lost	Won	Lost	Lost	Lost	Lost
Lost	Lost	Won	Lost	Lost	Lost	Lost	Won
Won	Lost	Lost	Lost	Lost	Lost	Lost	Lost
Lost	Lost	Lost	Lost	Lost	Won	Lost	Lost
Lost	Lost	Lost	Lost	Lost	Lost	Lost	Lost
Won	Lost	Lost	Won	Lost	Lost	Lost	Won

Male or Female Photos

An internet marketer is testing two different approaches to engaging a customer. Website visitors in group M were shown the product held by a male model. Website visitors in group F were shown the product held by a female model. The success in sales was tracked as "Bought" or "Left".

Use the IF(), SUM(), and COUNT() functions in excel to complete this exercise.

1. What kind of data was collected: Nominal, Ordinal, Interval, or Ratio?
2. What is the win ratio for Group M? Group F? (Win Ratios are calculated by finding the percentages of individuals that bought in relation to all possible events).
3. Create a 100% stacked bar chart that shows the win and loss frequencies for each group on a single graph.
4. What is the overall win ratio across both groups?
5. Based on this data alone, would you suggest that the internet marketer should use a male or female model? How confident are you in your answer?

Group M				Group F			
Left	Left	Left	Left	Bought	Left	Left	Left
Left	Left	Left	Bought	Left	Left	Left	Left
Left	Left	Left	Bought	Left	Left	Left	Left
Left	Left	Left	Left	Left	Left	Left	Left
Left	Left	Left	Left	Left	Left	Left	Left
Left	Bought	Left	Left	Left	Left	Left	Left
Left	Left	Left	Left	Left	Left	Left	Left
Left	Left	Left	Left	Left	Left	Left	Bought
Left	Left	Left	Left	Left	Left	Left	Left
Left	Left	Left	Left	Left	Left	Bought	Left
Left	Left	Left	Left	Left	Left	Left	Left
Left	Left	Left	Left	Bought	Left	Left	Left
Left	Left	Left	Left	Left	Left	Left	Left
Left	Left	Left	Left	Left	Left	Left	Left
Left	Left	Left	Left	Left	Left	Left	Left
Left	Left	Left	Left	Left	Left	Left	Left
Bought	Left	Left	Left	Left	Left	Left	Left
Left	Bought	Left	Left	Left	Left	Left	Left

BAGELS

A researcher wants to know, on average, how often bagels are consumed. He surveys individuals asking:

How often do you eat bagels?

- [ ] Two or more times a week
- [ ] Weekly
- [ ] Monthly
- [ ] Never

The research collects the following data. Use the "Sort" tool and LOOKUP() and COUNTIF() functions in Excel to complete this exercise.

1. What kind of data was collected: Nominal, Ordinal, Interval, or Ratio?
2. How many responded in each category?
3. Create a cross tab table that reflects the response item, frequency, and probability.
4. Create a 100% stacked bar chart that shows frequencies for each response.
5. Assume the following interpretation of the response items: Two or more times a week is equivalent to 104 bagels per year. Weekly is equivalent to 52 bagels per year. Monthly is equivalent to 12 bagels per year. Never is equivalent to 0 bagels per year. What is the weighted average category incidence for bagels of this sample per year?

Bagel Incidence			
Monthly	Monthly	Monthly	Monthly
Monthly	Weekly	Weekly	Weekly
Monthly	Monthly	Monthly	Monthly
Monthly	Monthly	Weekly	Weekly
Weekly	Two or more times a week	Two or more times a week	Weekly
Monthly	Monthly	Two or more times a week	Monthly
Weekly	Monthly	Weekly	Monthly
Monthly	Monthly	Monthly	Never
Monthly	Monthly	Weekly	
Monthly	Two or more times a week	Monthly	Monthly

Hello Kitty Split Sample

A survey was conducted among high school teenage boys and girls asking for their level of agreement with the following statement:

	Strongly Disagree	Disagree	Neither Agree nor Disagree	Agree	Strongly Agree
I like Hello Kitty.					

After conducting the survey, the research collects the following data. Use the "Sort" tool and LOOKUP() and COUNTIF() functions in Excel to complete this exercise.

1. What kind of data was collected: Nominal, Ordinal, Interval, or Ratio?
2. Code the response items with the following encoding: Strongly Disagree is 1, Disagree is 2, Neither Agree nor Disagree is 3, Agree is 4, Strongly Agree is 5.
3. How many boys responded in each category? How many girls responded in each category?
4. Create a cross tab table that reflects the response item, frequency, and probability of both the boy and girl sample.
5. Create a 100% stacked bar chart that shows frequencies for each response for both boys and girls on the same plot.
6. On a 1 to 5 scale, with 5 being "Strongly Agree":
    a. What is the weighted average response for teenage boys?
    b. What is the weighted average response for teenage girls?
    c. What is the mean of the teenage boy and girl means?

Girls	Boys
Neither Agree nor Disagree	Disagree
Neither Agree nor Disagree	Neither Agree nor Disagree
Neither Agree nor Disagree	Agree
Neither Agree nor Disagree	Neither Agree nor Disagree
Strongly Agree	Agree
Disagree	Neither Agree nor Disagree
Strongly Agree	Strongly Disagree
Neither Agree nor Disagree	Neither Agree nor Disagree
Agree	Neither Agree nor Disagree
Strongly Agree	Disagree
Neither Agree nor Disagree	Neither Agree nor Disagree
Agree	Strongly Disagree
Strongly Agree	Neither Agree nor Disagree
Agree	Disagree
Agree	Neither Agree nor Disagree
Agree	Strongly Disagree
Strongly Agree	Disagree
Agree	Strongly Disagree
Agree	Strongly Disagree
Agree	Strongly Disagree
Agree	Neither Agree nor Disagree
Agree	Agree

Starbucks

A researcher hypothesized that older persons were not as good customers for Starbucks as were younger persons. These data are shown below.

	Respondents	
Purchase Frequency	Under 25	25 years old or older
Once a week or more	40	7
Once a month to once a week	51	32
Once a year or less	8	54

1. Calculate the response probability for each sample group and each response item.
2. Create a 100% stacked bar chart of the data that has both the under 25 and 25 or older data.
3. Assume the following interpretation of the response items: Once a week or more is equivalent to 104 purchases per year. Once a month to once a week is equivalent to 12 coffees per year. Once a year or less is equivalent to 1 coffee per year.
    a. What is the probability weighted average number of coffees purchased per year among respondents under 25?
    b. What is the probability weighted average number of coffees purchased per year among respondents 25 years old or older?
    c. What is the mean of the under and over 25 year old mean purchase frequencies?
4. Assume the average coffee sold at Starbucks is priced at $2.75 and has an 80% contribution margin.
    a. What is the profit earned per coffee?
    b. What is the annual profit earned on average from a respondent under 25 years old?
    c. What is the annual profit earned on average from a respondent 25 years old or older?
    d. What is the annual profit earned on average from a respondent under or over 25 years old, that is, over both groups?
5. Which market segment looks to be more profitable?

Plastic Injection Molding Firm Size

Across four metropolitan areas, the size of plastic injection molding companies was evaluated according to the number of employees. What is the mean of the means for these samples?

Sample	Sample Size	Sample Mean
Chicago	32	52
Milwaukee	13	14
Detroit	28	70
Minneapolis	14	37

Firm Performance

Sales performance data over 26 weeks for a firm were collected.

Even or Odds

1. What is the average number of sales in an even week?
2. What is the average number of sales in an odd week?
3. What is the mean of means between even and odd weeks?
4. What is the average of the lowest selling 6 weeks?
5. What is the average of the highest selling 6 weeks?
6. What is the mean of means between the lowest and highest selling 6 weeks?

Profit

7. If each unit was priced at $14.99, and the cost of production is $2.75:
    a. What is the total channel contribution margin?
    b. What is the total channel markup?
8. If the manufacture produces an item for $2.75, sells that item for $7.50 to a retailer, and the retailer subsequently marks up the price by 100%:
    a. What is the manufacturer's contribution margin?
    b. What is the price at which the retailer sells the product?
    c. What is the retailer's contribution margin?
9. What is the total contribution margin of the channel (profit contribution from the retailer and manufacturer combined as a percentage of the end customer's price)?

Week	Unit Sales	Week	Unit Sales
1	10,662	14	14,180
2	15,119	15	15,398
3	16,751	16	19,882
4	13,493	17	15,951
5	9,573	18	14,732
6	14,284	19	28,056
7	15,244	20	15,375
8	22,138	21	16,843
9	34,342	22	12,795
10	18,446	23	13,131
11	11,544	24	11,004
12	11,940	25	14,206
13	10,162	26	10,440

CHAPTER 4: NORMAL DISTRIBUTIONS, Z-SCORES, AND CONFIDENCE INTERVALS

As we discovered with numerical data in Chapter 2, we can describe the distribution of measurements with a central tendency, sample standard deviation, skewness, and kurtosis. In several instances where the skewness was relatively mild, the plot of the frequency distribution followed a pattern that can well be described as a normal distribution.

The normal distribution describes a wide variety of population distributions. As such, we will broaden our examination to include not only a specific sample's responses, but an entire population's responses. Doing so will allow us to describe specific responses as well as entire distributions in general terms. Z-scores are a generalized method for comparing specific responses or groups of responses to population means. The standard normal distribution is a generalized method for describing distributions of responses.

In this chapter, students will learn:

- The difference between population means and population standard deviations, and sample means and sample standard deviations.
- How frequency distributions can be described with the normal distribution.
- How Z-scores measure the deviation of a data point or sample mean from the population mean in a standardized manner.
- How confidence intervals can be used to describe the expectation of finding data at a given point within a distribution.
- The meaning of an outlier.
- How measures of the standard error of a sample mean can be used to determine if a sample is representative of the population as a whole.
- The use of the profit equation of the firm for conducting breakeven and scenario analysis.

POPULATION STANDARD DEVIATION

In exploring the mean of sample means in Chapter 3, we drew attention to the fact that if we sample the population to measure an item, and then repeat the sampling of the same population and re-measure the item, we might uncover slightly different results between the two measurements. Only by measuring the entire population's response to an item can we claim to have measured the population's response.

When we measure the population's response with numerical data, we can find the population's average score and the standard deviation of responses within the population.

We distinguish the population's measure from a sample's measure with the use of Greek letters. Hence, the population's mean is denoted by μ, the lower case Greek letter pronounced "mu", and the population's standard deviation is denoted by σ, the lower case Greek letter pronounced "sigma".

While the equation for the population's average remains unchanged, that for the standard deviation differs slightly. Specifically, the denominator within the square roots of the population's standard deviation simply uses the size of the population. There is no "-1". We had to subtract one from the count of data elements when finding the sample's standard deviation due to the use of one degree of freedom in calculating the average. When we are measuring the entire population, we do not have to subtract this degree of freedom.

Hence, the population arithmetic average is given by

$$\mu = \frac{\sum_{i=1}^{N} X_i}{N}$$

and the population's standard deviation is given by

$$\sigma = \sqrt{\frac{\sum_{i=1}^{N}(X_i - \mu)^2}{N}}$$

In excel, σ can be calculated similarly to s but with the use of the function STDEV.P().

Normal Distribution

In examining frequency distributions, we found that they often peak about a central value and trail off for higher or lower values. We uncover this distribution pattern in a wide variety of numerically scored measures. In fact, it is so common that we have formalized our understanding of this type of distribution and have defined it to be the normal distribution.

The normal distribution, also describe as a bell curve, Gaussian curve, and normal curve, can be formally described as

$$P(x) = \frac{1}{\sqrt{2\pi\sigma^2}} e^{\frac{-(x-\mu)^2}{2\sigma^2}}$$

Where P(x) is the probability of finding the value x, μ is the population's mean, σ is the population's standard deviation, π is the mathematical constant approximately equal to 3.14, and e is the natural number approximately equal to 2.72.

Note that to find the probability at a point we must have three pieces of information: the point to be investigated, the mean, and the standard deviation. Where possible, researchers will use the population's mean and standard deviation. At times, the population's mean and standard deviation are unavailable. Sometimes researchers will use the sample's mean and standard deviation to approximate the normal probability at a point. Other times, researchers may use a different measure of the sample's central tendency to calculate the normal probability. And at still other times, a different metric of the mean and standard deviation will be devised. Which is used is dependent on the nature of the question to be investigated, and the nature of the data available.

Normal Probability Distributions and Excel

With Excel, we can find the normal probability at a point using the function NORMDIST(x, mean, standard deviation, FALSE). We will demonstrate why we need the fourth argument in this normal distribution function, currently set at "FALSE", in the next chapter. For now, take it as an answer to the question "Do you want the cumulative distribution function?", answer: "FALSE" for NO.

Notice that, apart from this, the Excel function requires the same three arguments: the point to be investigated, the mean, and the standard deviation.

To gain familiarity with the normal distribution, let us create one in Excel. A researcher measures an item on a population and identifies the mean at 50 and the standard deviation at 5. To calculate and plot the frequency distribution of scores, the researcher opens a blank worksheet in Excel.

- Labeling data as a good researcher will, she names Cell A1 "Mean" and Cell A2 "Standard Deviation", then inputs "50" in cell B1 and "5" in cell B2.
- She also labels column D with "Measurement" in cell D1 and column F with "Probability" in cell F1. Column D will contain the measurement or points for which the normal distribution will be calculated. Column F will contain the expected probability of a measurement returning that specific point.
- The researcher wants to be sure that she covers the range of possible measurements in her calculations and plotting. Since the average measurement is at 50 and the standard deviation is at 5, she doesn't expect she will get many scores below 0, nor many above 100. Hence, she decides to calculate the probability of points between 0 and 100. In cell D2 she types in 0. Then, in cell D3, she types in "=D2+1" in order to start generating a list of points to be investigated. She copies and pastes cell D3 down column D until she reaches the last point she wants to investigate, 100.
- Now, she is ready to calculate the probability at a point according to the normal distribution.
- In cell F2, she types "=NORMDIST(D2,B$1,B$2,FALSE)". The arguments of this function were identified with the following logic.
    - Cell D2 contains the point to be investigated. The function will return the normal probability of finding a measure at the point contained in cell D2, "0".
    - Cell B1 contains the mean to use in calculating the normal probability, 50. She puts a dollar sign in front of the row number to enable the cell's formula to be copied and pasted throughout the column while using the exact same mean for each calculation.
    - Cell B3 contains the standard deviation to use in calculating the normal probability, 5. Again, she puts a dollar sign in front of the row number to enable the cell's formula to be copied and pasted throughout the column while using the exact same standard deviation for each calculation.
    - She types in "FALSE" for the function's fourth argument.
- She then copies and pastes cell F3 down column D until she reaches the last point she wants to investigate, 100.
    - Note that some of the calculations may look like "1.53892E-23". This is simply scientific notation. It means 1.53892 X 10^{-23}, or 0.0...0153892 where the one starts 23 digits after the decimal point. We usually don't, however, work with scientific notation in marketing. You may want to change the cell's format to read as a normal decimal number. Use the "Number Format" panel on the "Home" tab to adjust the format as you see fit.
- The results should appear similar to that shown in the table below.
- She also plots this data. The normal distribution function is a continuous function, and, as such, a simple line plot would best represent the function.
    - In a blank cell, insert a simple line chart from the "Insert" tab in Excel.
    - Hit the "Select Data" button. Add a series using the series name "Probability" and the series values found in column D, those of the measurement's probability. Edit the Horizontal Axis Labels to include the range of cells in column D, the points for which the probabilities were calculated.

- Delete the legend (there is only one series of data being plotted, hence the legend is unnecessary in this plot), add appropriate Axis Titles, and title the graph "Normal Probability Distribution μ=50, σ =5". (Use the symbol font to type Greek letters.)

Measurement	Probability
0	0.00
1	0.00
2	0.00
⋮	⋮
37	0.00
38	0.00
39	0.01
40	0.01
41	0.02
42	0.02
43	0.03
44	0.04
45	0.05
46	0.06
47	0.07
48	0.07
49	0.08
50	0.08
51	0.08
52	0.07
53	0.07
54	0.06
55	0.05
56	0.04
57	0.03
58	0.02
59	0.02
60	0.01
61	0.01
62	0.00
63	0.00
⋮	⋮
100	0.00

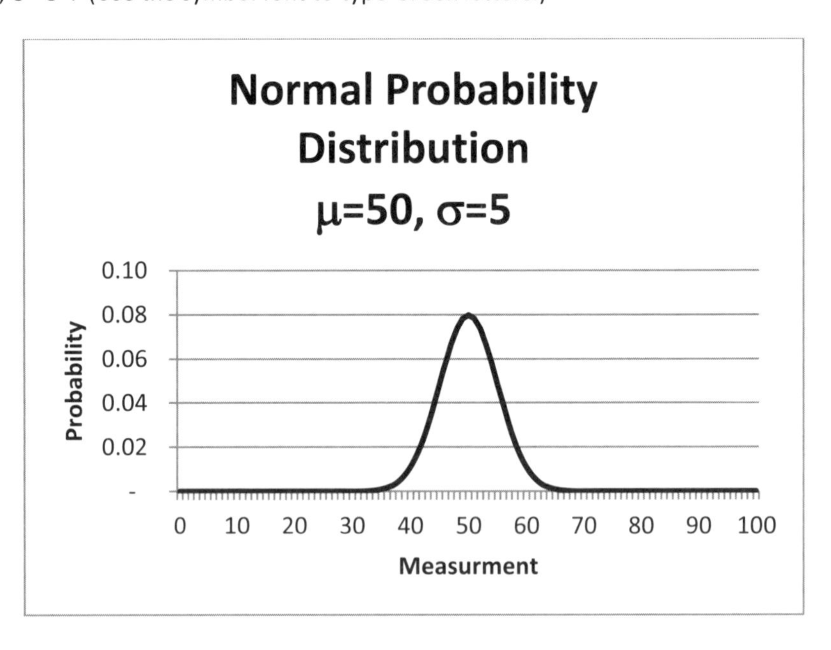

Attributes of Normal Probability Distribution

Notice in this frequency distribution plot of the normal probability distribution the following attributes regarding its shape:

1. The distribution is unskewed. Normal distributions have exactly zero skewness .
2. The distribution peaks at 50, the average value. Normal distributions will always peak at the average value. If the average increases, the peak will shift upwards to higher measurements. If the average decreases, the peak will shift downwards to lower measurements. For instance, compare the normal probability distributions with means of 35, 50, and 75. (A legend is used in this plot to identify the three different means).

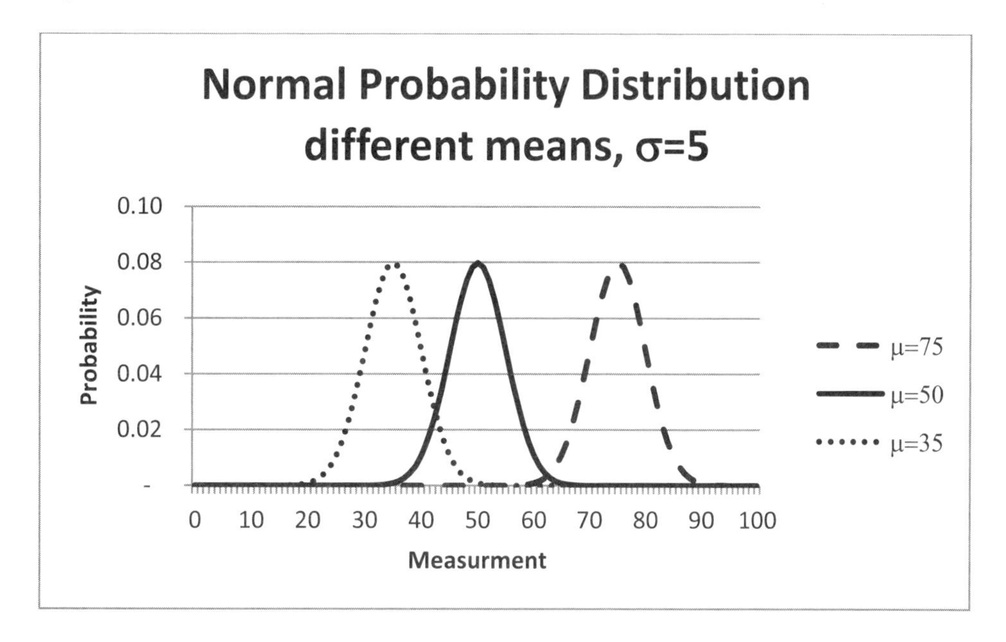

3. For a mean of 50 and a standard deviation of 5, the distribution of measures has probabilities roughly equal to zero below 30 and above 70. The more probable measures are actually found in a limited range between 40 and 60, or plus or minus two standard deviations where 2X5 = 10. This is a general property of normal distributions. If the standard deviation was larger, the normal probability distribution would be broader. If it was smaller, the normal probability distribution would be narrower. For instance, compare the normal probability distributions with standard deviations of 2, 5, and 15.

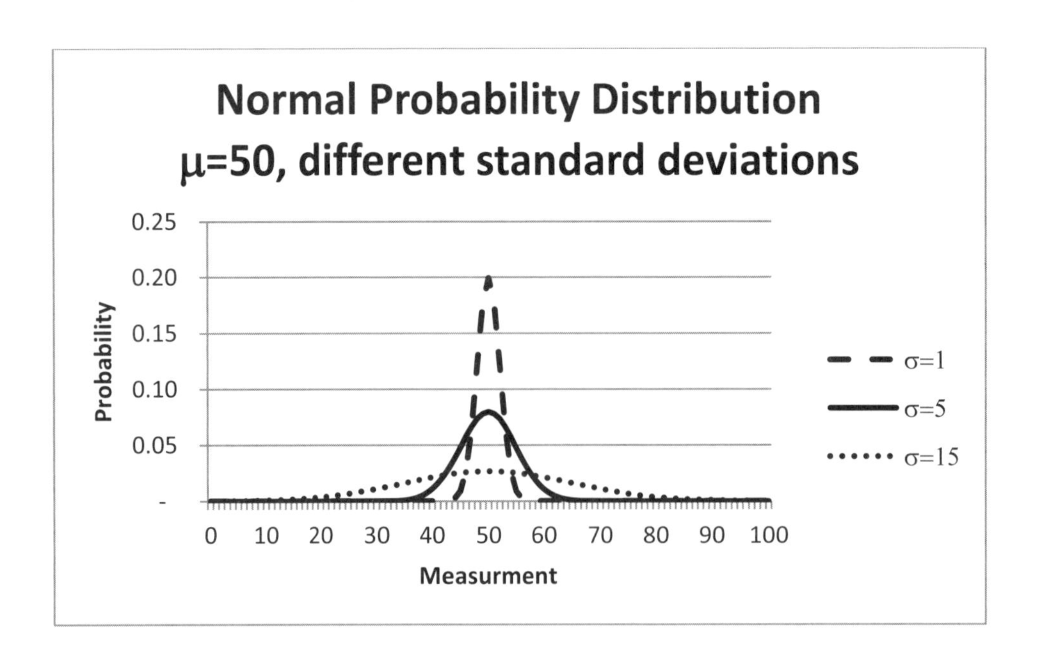

Standardized Scores

If we know the population's mean and standard deviation, we can standardize scores. **The standardized score, also known as the normalized score or Z-Score, describes how far away a score is from the average. More precisely, it measures the number of standard deviations a given score is away from the mean.**

$$Z = \frac{X_i - \mu}{\sigma}$$

Z-scores are very useful for comparing data across samples and measurements precisely because of their standardized nature. As such, researchers will often think more in terms of Z scores than actual measurement results. Students of quantitative marketing are well advised to gain a natural familiarity with Z-scores. After reflecting on the following descriptions, you should be able to describe the meaning of a Z score equal to 0.5.

- A Z score of **0** implies that a particular response **matched the population's average** response.
- A Z score of **1** implies that a particular measurement was **one standard deviation above** the population's mean.
- A Z score of **-1** implies that a particular measurement was **one standard deviation below** the population's mean.
- A Z score of **2** implies that a particular measurement was **two standard deviations above** the population's mean.
- A Z score of **-2** implies that a particular measurement was **two standard deviations below** the population's mean.
- A Z score of **.75** implies that particular measurement was **.75 standard deviations above** the population's mean.
- A Z score of **-3.15** implies that particular measurement was **3.15 standard deviations below** the population's mean.

When comparing measurements across survey items, different surveys, or different tests, Z scores enable a consistent method to explore issues of

- How different is this measurement compared to that measurement?
- In what ways is this respondent, or group of respondents, different from another group of respondents?
- Is this response an outlier, that is, is it truly different from the average response, or is it relatively similar to the other responses?
- And, importantly for determining if a response is statistically interesting at all, is this measurement statistically different from zero?

For example, we can use standardized scores to describe the results of an intelligence quotient. The population mean IQ is 100 and the population standard deviation is 15. Thus

- A person with an IQ of 110 has a Z score of .67, not very different from the average and definitely within a standard deviation.
- A person with an IQ of 95 has a Z score of -.33, not very different from the average and definitely within a standard deviation.
- A person with an IQ of 115 has a Z score of 1, somewhat higher than the average at one standard deviation.
- A person with an IQ of 69 has a Z score of -2.1, definitely lower than the average at more than two standard deviations. Such a person would be qualified as a "special needs" individual by most US governmental agencies and would receive extra governmental support.
- A person with an IQ of 132 has a Z score of 2.13, definitely higher than the average at more than two standard deviations. Such a person would qualify to join Mensa International.

In general, to interpret a Z score, one could claim that

- Measurements that lie within 1 standard deviation of the mean, or have a Z score between -1 and 1, are relatively average measurements.
- Measurements that lie within 2 standard deviations of the mean, or have a Z score between -2 and 2, are also relatively normal measurements.
- Measurements that lie outside of 2 standard deviations of the mean, or have a Z score less than -2 or greater than +2, are definitely different from the average. These scores may warrant closer attention, or may be thought of as outliers or as indicators that a measurement can be considered to be different from zero.

Standard Normal Distributions

Because normal distributions are so common, we often create the standard normal distribution. The standard normal curve has a mean of 0 and a standard deviation of 1. With this mean and standard deviation, the normal distribution can be written simply as

$$P(Z) = \frac{1}{\sqrt{2\pi}} e^{\frac{-z^2}{2}}$$

We use a Z with the standard normal distribution to indicate the use of standardized scores.

With Excel, we can find the normal probability for a given standardized score using the =NORMDIST(Z,0,1,FALSE), where Z is the standardized score under investigation, and the mean and standard deviation of the standard normal distribution is zero and one respectively.

CONFIDENCE INTERVALS

As we see in the normal probability distribution plots, a measure is far more likely to be near the population mean than far from the population mean. Moreover, the probability of finding a measure different from the population mean is dependent on the population standard deviation of the measure. A measure 10 units away from the population mean is highly unlikely if the standard deviation is 1, but can be somewhat expected if the standard deviation is 20.

Confidence intervals define a range of values in which we would expect to find a measurement. More precisely, it defines an interval about a population mean in which we can be confident, at some predetermined level, that a measure from a similar sample will yield a mean within that interval. The predetermined level is called the confidence level. The size of a confidence interval is dependent on the level of confidence we wish to have of finding a measurement within the interval.

If we want it to be highly likely that a measure from the same population will be found within the confidence interval, we would choose a high or somewhat high confidence level.

- With a confidence interval defined at the 90% confidence level, we would expect a measure from the same population to be found within this confidence interval 90 out of 100 times, or, equivalently, 9 out of 10 times.
- With a confidence interval defined at the 95% confidence level, we would expect a measure from the same population to be found within this confidence interval 95 out of 100 times, or, equivalently, 19 out of 20 times.
- With a confidence interval defined at the 99% confidence level, we would expect a measure from the same population to be found within this confidence interval 99 out of 100 times.

If we want it to be somewhat less likely that a measure from the same population will be found within the confidence interval, we would choose a smaller confidence level.

- With a confidence interval defined at the 20% confidence level, we would expect a measure from the same population to be found only 20 out of 100 times, or, equivalently, 1 out of 5 times.
- With a confidence interval defined at the 50% confidence level, we would expect a measure from the same population to be found within this confidence interval 50 out of 100 times, or, equivalently, 1 out of 2 times.
- With a confidence interval defined at the 68% confidence level, we would expect a measure from the same population to be found within this confidence interval 68 out of 100 times.

For a normal probability distribution, the size of the confidence interval associated with a given confidence level is well known. See the figure below.

- The 68% confidence interval, a rather narrow interval, covers the range of measures spanning one single standard deviation above and below the mean. A majority of measures will lie within a single standard deviation of the mean, but several will lie outside of this mean.
- The 95% confidence interval, a somewhat broader interval, covers the range of measures spanning 1.96 standard deviations above and below the mean. 19 out of 20 measures will lie within approximately two standard deviations of the mean. The probability of finding a measure from the same population outside the range defined by the mean plus or minus 1.96 standard deviations is relatively small, at only a 1 in 20 chance.
- The 99% confidence interval, an even broader interval, covers the range of measures spanning 2.58 standard deviations above and below the mean. 99 out of 100 measures will lie within 2.58 standard deviations of the mean. The probability of finding a measure from the same population outside the confidence interval defined at the 99% confidence level is extremely small, at only a 1 in 100 chance.

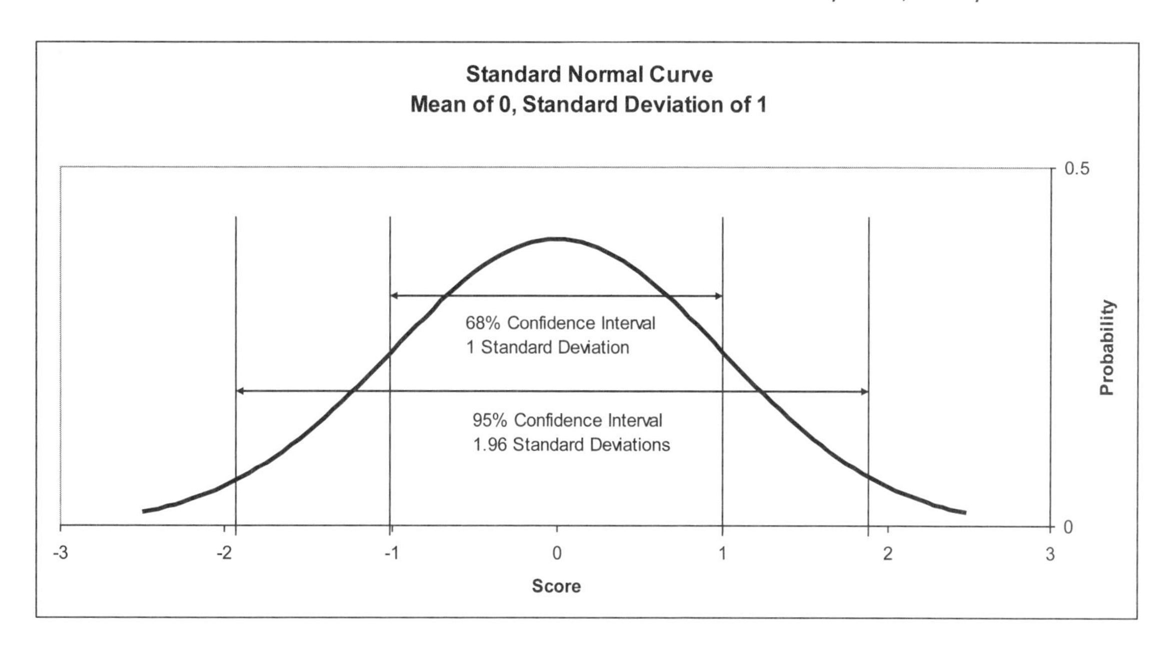

In general, the probability of finding a measure within any range can be determined. **A quantitative analyst should know the following metrics.**

Description		Interval Lower Limit	Interval Upper Limit	Probability
Within a half a standard deviation of the mean	38% confidence interval	$\mu - 0.5\sigma$	$\mu + 0.5\sigma$	38%
Within a standard deviation of the mean	68% confidence interval	$\mu - 1\sigma$	$\mu + 1\sigma$	68%
	90% confidence interval	$\mu - 1.65\sigma$	$\mu + 1.65\sigma$	90%
Within roughly 2 standard deviations of the mean	95% confidence interval	$\mu - 1.96\sigma$	$\mu + 1.96\sigma$	95%
	99% confidence interval	$\mu - 2.58\sigma$	$\mu + 2.58\sigma$	99%
Anywhere below the mean		$-\infty$	μ	50%
Anywhere above the mean		μ	$+\infty$	50%
Between the mean and 1 standard deviation above the mean		μ	$\mu + 1\sigma$	34%
Below 1 standard deviation above the mean		$-\infty$	$\mu + 1\sigma$	84%
The bottom 90 percent		$-\infty$	$\mu + 1.28\sigma$	90%
The top 10 percent		$\mu + 1.28\sigma$	$+\infty$	10%
The top 5 percent		$\mu + 1.65\sigma$	$+\infty$	5%
The top 2 percent		$\mu + 2.06\sigma$	$+\infty$	2%
The top 1 percent		$\mu + 2.33\sigma$	$+\infty$	1%

Measurements and Expectations

Z Scores, the Normal Probability Distribution, and Confidence Intervals

As the above discussions indicate, Z scores describe more than simply the deviations, they also describe the likelihood of finding a measure at a point when the data is normally distributed. Consider the following few Z scores and the normal distribution.

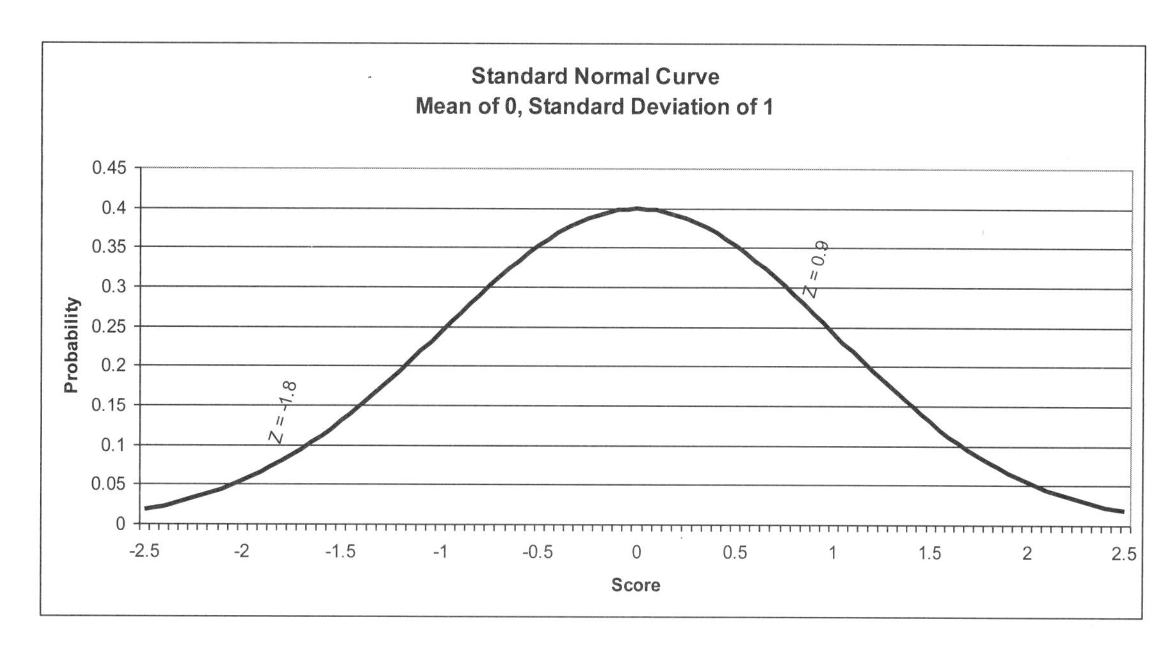

A measure with a Z score of 0.9 or -1.8 is relatively common within a normally distributed set of measures. Both of these scores can be found within the 95% confidence interval, where any measure of the sample would yield a Z score of between +/- 1.96 by chance alone.

However, a Z score outside of the 95% confidence interval, at 3.2 perhaps, is truly unique, and would imply that the measure from that sample differs from the expected measure of the population as a whole, or, more precisely, that a Z score of 3.2 is highly unlikely from the population as a whole. Differences like these, when they are related to market segments, can imply that a particular market segment is unique, perhaps in the products or services it demands, or in its willingness to pay.

Outliers

Outliers are data that lie outside of the expected range. Generally, we use the 95% confidence interval to identify outliers. Normally distributed measures with a Z score outside of the range of +/- 1.96 will occur from the same population only 5% of the time or less. In other words, only 1 in 20 times will a researcher find a value outside the 95% confidence interval. As such, samples with measures outside of the 95% confidence interval are rare. When the measure comes from a specific single measure, we call that data-point an outlier.

Outliers are usually of statistical interest and deserve greater attention.

- **At times, outliers indicate a bad piece of data and just have to be discarded.**
- **At other times, they indicate that that data is highly skewed and the median will be a better representation of the central tendency.**

- **At still other times, outliers indicate that a measure was truly interesting and different from the mean, perhaps due to an underlying difference between the sample and the average population. When the latter occurs, the researcher may have uncovered a new way to examine the market. Uncovering significant differences between samples enables marketers to better direct their efforts.**

Because the status of being an outlier alone doesn't indicate a definitive conclusion, researchers should examine other sources of data to determine how to handle an outlier and if it is truly novel.

Sample Means and Standard Errors

The confidence interval of a frequency distribution of an entire population describes the likelihood of getting a single measurement within a range, as described above. At times, though, a researcher will know attributes of the population as a whole as well as the mean measure of a sample, and will desire to know if the sample is representative of the population. This requires calculating the standard error of a sample measurement.

Different samples from the same population will have different means. Simply because the sample mean is different from the population mean does not imply that the sample is not representative of the population as a whole. Rather, random sampling errors will result in different samples having slightly different means.

The standard error of the sample mean is a measure of potential fluctuations of the mean arising from random sampling fluctuations. It is calculated by dividing the population standard deviation by the square root of the sample size and is often denoted by SE. With a population standard deviation given by σ and a sample size given by N, the standard error, SE, is found with the following equation.

$$\boldsymbol{SE = \frac{\sigma}{\sqrt{N}}}$$

With the standard error as defined above, a quantitative analyst can determine whether a sample is representative of the population or not by calculating the confidence interval of the sample mean and comparing it to the population's mean. For instance, consider a sample of 25 measurements with a mean measure of 2.45 when compared with a population with a mean measure of 2.30 and a standard deviation of 0.5. Given this information, a quantitative analyst can calculate the confidence interval about the sample mean.

1. First, the analyst calculates the standard error of the measure:
$$SE = \frac{\sigma}{\sqrt{N}} = \frac{0.5}{\sqrt{25}} = 0.1$$
2. Second, the analyst determines the confidence level she wishes to use. Keeping with convention, she chooses the 95% confidence level, which implies that the confidence interval will range from 1.96 standard errors below the mean to 1.96 standard errors above the mean.
$$\text{Confidence Interval (CI)} = \text{Sample Mean } \pm 1.96 \text{ Standard Errors}$$
$$CI = \bar{X} \pm 1.96\,SE$$
$$CI = 2.45 \pm 1.96 \cdot 0.1$$
$$CI = 2.45 \pm 0.196$$
$$CI = [2.25, 2.65]$$
3. Finally, the analyst compares the 95% confidence interval of the sample mean to the population mean. The population mean in this instance was 2.30. This mean is contained within the 95% confidence interval of

the sample mean. Hence, the analyst can conclude that the sample is representative of the population as a whole.

MARKETING METRICS: BREAKEVEN & SCENARIO ANALYSIS

Understanding the profit equation of the firm enables marketers to uncover the requirements for business survival, consider the effects of altering specific elements of the marketing mix on the health of the firm, or even positing hypothetical questions and uncovering their impact.

A breakeven analysis demonstrates the required sales volume for business survival. Breakeven analysis can include a breakeven plot or the identification of the breakeven volume.

BREAKEVEN PLOTS

A breakeven plot is a plot of the profit or loss of the firm as sales volume increases, given a defined price, variable cost, and fixed cost. For instance, consider a firm selling French fries for $1.50, with a variable cost of $0.25 and a monthly fixed cost of $2000. An analyst will create a breakeven plot in Excel with the following steps.

- Define and label cells for "Price", "Variable Cost", and "Fixed Cost".
- Define columns for Quantity Sold and Profit.
- In the Quantity Sold column, they will choose a starting point, say zero in this case, and increment the sales volume by a fixed amount for each cell going down, say 100 units in this case.
- In the Profit column, they will use the profit equation of the firm, $\pi = Q \bullet (P\text{-}V) - F$, where π represents profit, to calculate the profit at each associated quantity sold. Excel will allow them to use the previously defined cells for price, variable cost, and fixed cost rather than typing in a fixed number. The analyst chooses this approach so that she can also change these parameters at will for a future calculation.
- Once the line has been defined from the firm's parameters and the profit equation of the firm, the analyst creates a breakeven plot by inserting a line plot of profit versus volume.
- The results might look like that shown below. From the breakeven plot, we see the firm suffers heavy losses for sales volumes less than 1500 units, and begins to turn a profit only after clearing nearly 1600 units.

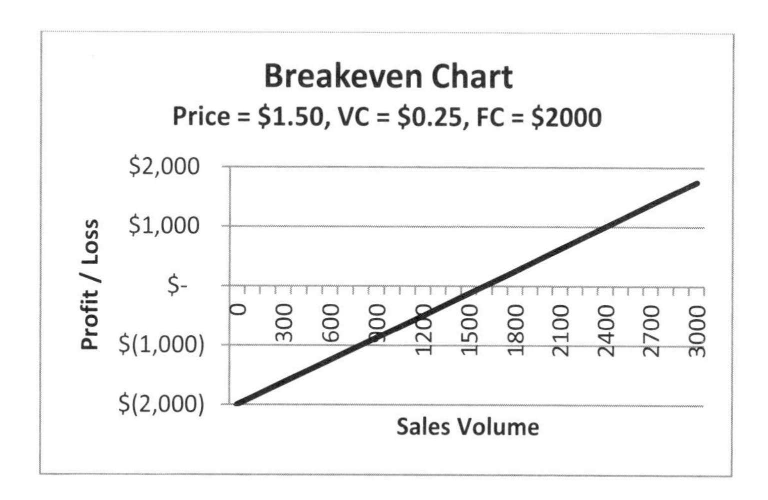

	A	B	C	D	E
1	Price	$ 1.50		Quantity Sold	Profit
2	Variable Cost	$ 0.25		0	=D2*(B$1-B$2)-B$3
3	Fixed Cost	$ 2,000		100	$ (1,875.00)
4				=D3+100	$ (1,750.00)
5				300	$ (1,625.00)
6				400	$ (1,500.00)
7				500	$ (1,375.00)
8				600	$ (1,250.00)
9				700	$ (1,125.00)
10				800	$ (1,000.00)
11				900	$ (875.00)
12				1000	$ (750.00)
13				1100	$ (625.00)
14				1200	$ (500.00)
15				1300	$ (375.00)
16				1400	$ (250.00)
17				1500	$ (125.00)
18				1600	$ -
19				1700	$ 125.00
2				1800	$ 250.00
20				1900	$ 375.00
21				2000	$ 500.00
22				2100	$ 625.00
23				2200	$ 750.00
24				2300	$ 875.00
25				2400	$ 1,000.00
26				⋮	⋮

Breakeven Volumes

While the breakeven plot provides a graphic depiction of the required sales to generate a profit, a breakeven volume analysis is more direct. It identifies the specific sales volume which the firm must achieve in order to generate profits. **Any sales volume below the breakeven volume results in a loss, and any sales volume above the breakeven volume results in a profit. A sales volume equal to the breakeven volume implies the firm neither generates profits nor suffers losses, but simply breaks even.** The breakeven volume is also sometimes called a volume hurdle, or the volume which the sales and marketing team must overcome in order to sustain the firm.

Because the breakeven volume is defined by the sales volume at which the firm's profits are zero, we can identify the breakeven volume from the profit equation of the firm.

Given

Profit = Q •(P– VC)– FC

at the breakeven volume, we know that profit is zero. If Profit = 0, the breakeven volume Q_{BE} can be derived as

$0 = Q_{BE} \bullet (P - VC) - FC$

Rearrange the equation to find

$$FC = Q_{BE} \cdot (P - VC)$$

$$\frac{FC}{(P - VC)} = Q_{BE}$$

or

$$\boldsymbol{Q_{BE} = \frac{FC}{(P - VC)}}$$

This is known as the breakeven volume. For our French fry vendor with a price of $1.50, variable costs of $0.25, and a monthly fixed cost of $2000, the breakeven volume is 1600 units.

SCENARIO ANALYSIS

A scenario analysis follows from an "If this, then what?" line of questioning. For instance, "If our French fry vendor increased sales by 100 units, what would that imply for profits?" The more general scenario analysis types of questions usually refrain from asking about a specific fixed size parameter change, but rather focus on a percentage increase or decrease of a parameter that defines the firm's profits.

For an example of a strong scenario analysis, we could consider a 10% improvement analysis. In a 10% improvement analysis, we identify the profit impact of improving any specific characteristic of the firm by 10% above the current status quo. Building on our French fry example, let us assume that last month's sales were 2234 units at $1.50 with variable costs of $0.25 and monthly fixed of $2000. The 10% improvement test might ask the following questions:

1. If price went up 10%, keeping everything else constant, what would happen to profits?
2. If sales volume (quantity sold) went up 10%, keeping everything else constant, what would happen to profits?
3. If variable costs went down 10%, keeping everything else constant, what would happen to profits?
4. If fixed costs went down 10%, keeping everything else constant, what would happen to profits?

Each of these questions is well posed for analysis. Using subscripts of i for initial and f for final, we can reframe the question as follows:

- Each time the firm wants to know "what would happen to profits?", the analyst would calculate the new profitability as well as the percent change in profits ($\%\Delta\pi$), where percent change in profits is calculated by comparing the change in profits calculated as the difference between the final profits (π_f) and the initial profits (π_i), to the starting profits.

$$\%\Delta\pi = \frac{\pi_f - \pi_i}{\pi_i}$$

- For 10% price and sales volume increases, the analyst calculates new prices and volumes as

$$P_f = P_i \cdot (1 + 10\%)$$
$$Q_f = Q_i \cdot (1 + 10\%)$$

- For 10% variable and fixed costs reductions, the analyst calculates new variable and fixed costs as

$$VC_f = VC_i \cdot (1 - 10\%)$$
$$FC_f = FC_i \cdot (1 - 10\%)$$

The reframing of these questions then allows the analyst to create the following table in Excel.

	Initial Condition	10% Increase in Price	10% Increase in Sales Volume	10% Decrease in Variable Cost	10% Decrease in Fixed Cost
Price	**$ 1.50**	**$ 1.65**	$ 1.50	$ 1.50	$ 1.50
Sales Volume	**2,234**	2,234	**2,457**	2,234	2,234
Variable Cost	**$ 0.25**	$ 0.25	$ 0.25	**$ 0.23**	$ 0.25
Fixed Cost	**$ 2,000**	$ 2,000	$ 2,000	$ 2,000	**$ 1,800**
Profit	**$ 792.50**	**$1,127.60**	**$1,071.75**	**$ 848.35**	**$ 992.50**
Change in Profit		$ 335.10	$ 279.25	$ 55.85	$ 200.00
% Change in Profit		**42%**	**35%**	**7%**	**25%**

The above scenario analysis is quite revealing.

- First, it indicates that the largest impact on profitability is achieved through price: a minor improvement in price has a major improvement on profits. A further analysis would reveal that a small degradation of prices has a large negative impact on profits. Because price is one of the marketing mix variables under the control of a marketer, and because many of the tactics marketers use to attract customers is price, strategic marketers will pay close attention to any action that impacts price, including sales promotions, coupons, price reductions, and price increases.
- The 10% improvement analysis also indicates that the sales volume has the second largest impact on profits. Marketers and salespeople are charged with driving sales volume through marketing communications, interactions with customers, and choices of distribution channels. The achievement of sales targets determines the health of the firm. Failure to achieve sales goals implies that the firm is not sufficiently relevant to its customers or is otherwise failing to attract them in sufficient numbers. Small improvements in sales volumes have disproportionately large positive impacts on profits and small losses of sales volumes have disproportionately large negative impacts on profits.
- A 10% improvement in variable costs may be achieved through improvements in operations or stronger bargaining with customers. From this analysis, we see that a variable cost reduction is important, but not as important as improving one of the other parameters within the profit equation of the firm.

- A 10% improvement in fixed costs may be achieved through reductions in overhead costs, often in the form of firing staff employees. From this analysis, we see that a fixed cost reduction has significant impact for the French fry operator. If sales volume or price cannot be improved, we should expect the French fry operator to seek to reduce fixed costs.

A scenario analysis can also be used to consider more complex questions. For instance, in pricing strategy, a price increase is often coupled with a volume decrease, and vice versa. Similarly, in product strategy, product improvements often increase variable costs but are carried out in order to increase sales volume as well. We can deploy the same approach to these more complex scenarios.

Let us quantify these questions as follows and calculate the percent change in profits to generate the following table.

1. If price went up 10% and volume decreased by 10% at the same time, what would happen to profits?
2. If price went down by 10% and volume increased by 10% at the same time, what would happen to profits?
3. If variable costs went up 10% and sales increased by 10% at the same time, what would happen to profits?

	Initial Condition	10% Increase in Price and 10% Decrease in Volume	10% Decrease in Price and 10% Increase in Volume	10% Increase in Variable Cost and Quantity Sold
Price	$ 1.50	$ 1.65	$ 1.35	$ 1.50
Sales Volume	2,234	2,011	2,457	2,457
Variable Cost	$ 0.25	$ 0.25	$ 0.25	$ 0.28
Fixed Cost	$ 2,000	$ 2,000	$ 2,000	$ 2,000
Profit	$ 792.50	$ 814.84	$ 703.14	$ 1,010.32
Change in Profit		$ 22.34	($ 89.36)	$ 217.82
% Change in Profit		3%	-11%	27%

From this more complex scenario analysis, we realize

- A price increase that is associated with a somewhat similar sales volume decrease is likely to improve profits. The same cannot be said for the inverse.
- An improvement of the product that delivers a proportionate improvement in sales volume is likely to greatly improve profits.

Other, more complex scenarios can be quantified and investigated similarly.

EXERCISES

DIFFERENCES? IT'S ALL NORMAL

A 1 to 5 Likert scale was used to measure attitudes regarding a brand. The researcher decides to compare their measurements to a normal distribution by plotting a normal distribution curve. Define the horizontal axis as starting at 1 and going to 5 with an increment of 0.1)

1. On the same chart, plot the normal distribution curves for
   a. A mean of 3.5 and a standard deviation of 0.7.
   b. A mean of 1.5 and a standard deviation of 0.7.
   c. A mean of 4.5 and a standard deviation of 0.7.
2. On the same chart, plot the normal distribution curves for
   a. A mean of 2.5 and a standard deviation of 0.7.
   b. A mean of 2.5 and a standard deviation of 2.3.
   c. A mean of 2.5 and a standard deviation of 0.4.

PAPER MANUFACTURING

A computer paper manufacturer has a production process that operates continuously throughout a production shift. The paper produced has an average length of 11 inches. The standard deviation is 0.02 inch. Every so often, samples are taken to determine whether the average paper length is still equal to 11 inches or whether something has gone wrong to change the length. A random sample consisting of 100 sheets showed the average (mean) to be 10.998 inches.

1. Calculate the standard error of the sample mean.
2. Construct a 95% confidence interval around the sample mean of 10.998 inches.
3. Is the population mean of 11 included in the interval?
4. If not, the sample mean differs significantly from the standard, which is the average length of 11 inches, i.e. something has gone wrong to change the length. If 11 is included in the interval, the sample mean does not differ significantly from it and the process is under control. Is the process under control or not?

FIFA – ¡VIVA ESPAÑA!

Grand Manufacturing produces FIFA soccer balls. Weights are supposed to be normally distributed with a mean of 15.094 ounces and a standard deviation of 0.273 ounces. A sample of 50 soccer balls shows a mean weight of 15.31 ounces.

1. Calculate the standard error of the sample mean.
2. Construct a 95% confidence interval around the sample mean.
3. Is the population mean included in the interval?
4. Are the soccer balls of an acceptable weight?

RADIOHEAD

Marissa has been selling Radiohead t-shirts in a department store for 5 years. After much positive feedback from her friends and family about her products, she decides to open up her own store with Radiohead products. Marissa knows that convincing a bank to help fund her new business will require more than a few positive testimonials from family. Based on her selling experience, Marissa believes that Radiohead fans in her area spend more than the national average on Radiohead t-shirts. This fact could make her business successful.

Marissa would like to support her belief with data to include in a business plan proposal that she would then use to obtain a small business loan. After conducting research, she learns that the national average spending by Radiohead fans on Radiohead products is $49 every 3 months.

Marissa takes a random sample of 25 Radiohead fans and finds the sample mean ($\bar{X}$) is $54 and the population standard deviation s is $9. Is demand for Radiohead products higher in her town?

LEO BURNETT

Suppose you select employees from Leo Burnett and the annual income of each is determined. The mean is $65,700 and the standard deviation is $3,500.

1. If Corey Smith's annual income is $62,200, what is her Z score?
2. If Craig Smith's annual income is $68,700, what is his Z score?
3. If Carlyle Smith's annual income is $65,150, what is his Z score?
4. If Carolyn Smith's annual income is $72,800, what is her Z score?

Making the Bacon

The following table lists the 2001 sales employees' incomes, which includes base salary and bonuses. The employees are identified by their sales territories.

1. Compute the mean, median, & mode.
2. What is the range of incomes (minimum and maximum)?
3. What is the population's standard deviation?
4. What is the Z score of each state?

State	Income
Alaska	52,876
Montana	32,896
N. Dakota	34,457
Hawaii	51,046
Wyoming	38,186
S. Dakota	35,202
Washington	45,310
Utah	45,654
Nebraska	37,864
Oregon	39,305
Colorado	46,738
Kansas	40,438
California	46,499
Arizona	38,537
Oklahoma	33,448
Idaho	37,117
New Mexico	53,096
Texas	39,120
Nevada	42,177

ALL SOULS GATHERED

The following data regarding the number of souls gathered at Hyde Park Union Church for a number of Sundays has been collected. This constitutes the population of measurements.

1. What is the population mean and population standard deviation of souls gathered?
2. For each Sunday, calculate the Z score.
    a. What is the average Z score?
    b. Where are the outliers?
    c. What might explain the outlier Sundays?
3. Use the 95% confidence level for the following questions.
    a. What is the confidence interval?
    b. How many Sundays have attendance which falls outside of this confidence level?
    c. What percentage of the reporting Sundays does not fit within the 95% confidence interval?
4. Use the 90% confidence level for the following questions.
    a. What is the confidence interval?
    b. How many Sundays have attendance which falls outside of this confidence level?
    c. What percentage of the reporting Sundays does not fit within the 90% confidence interval?

All Souls Gathered								
68	89	105	102	75	95	114	95	88
89	97	79	102	93	76	100	152	105
66	93	102	81	135	78	92	110	91
80	83	108	72	110	125	103	105	83
61	87	90	80	106	80	82	99	144
66	90	110	59	97	102	107	111	106
87	106	110	85	82	83	189	115	120
72	97	115	96	107	114	92	112	118
55	90	129	92	180	107	73	130	73
108	194	97	73	106	98	61	146	115
69	111	47	54	97	108	70	67	113
101	75	102	97	97	103	71	128	126
115	89	111	121	94	115	65	100	99
92	82	94	111	112	90	75	115	127
100	98	84	113	104	127	93	115	139
93	83	99	97	105	128	71	95	175
107	71	82	98	81	97	100	100	130
104	80	102	102	110	63	111	120	88
95	71	117	117	105	104	90	115	117
97	64	99	146	73	118	115	127	121
108	80	106	112	75	85	101	96	130
93	100	105	104	76	99	102	94	80
135	67	110	92	75	101	122	95	54
93	69	175	152	87	114	98	120	78
99	99	81	97	72	112	95	113	105
124	104	100	95	58	134	95	101	115
75	88	90	84	78	91	97	105	110
88	57	118	98	100	67	105	95	120

Retail Firm Performance

Over a number of weeks, the retailer ran a different promotion affecting the average selling price. Sales and price data were collected each week for one year.

1. Calculate the revenue earned for each week as well as the volume weighted average price achieved (total revenue divided by total volume, or the sum of the products of price and volume divided by the sum of the volume).
2. Calculate
    a. The sample's mean revenue per week.
    b. The sample's standard deviation in revenue per week.
    c. The Z score for each week's revenue.
    d. Identify the outliers.
    e. What might explain the outlier data?
3. If unit cost was $7.50 and weekly fixed cost was $19,000:
    a. Create a breakeven plot as sales volumes go from 0 to 6000 units using the volume weighted average price.
    b. Identify the breakeven volume using the volume weighted average price.
    c. What is the breakeven weekly volume at the manufactures suggested retail price of $14.99?
    d. What was the average profit or loss earned per week?
    e. If the retailer held the price constant at $14.99 while achieving the same sales volume, what would have been the average profit/loss generated per week?

Units	Price	Units	Price
4,198	$ 14.56	2,238	$ 13.94
4,515	$ 14.38	1,610	$ 14.46
3,668	$ 14.50	1,648	$ 14.65
3,698	$ 14.45	1,937	$ 14.53
3,828	$ 14.56	1,963	$ 14.50
4,293	$ 14.37	1,897	$ 14.62
3,466	$ 13.78	2,065	$ 14.69
3,225	$ 14.57	1,973	$ 14.62
3,326	$ 14.40	2,216	$ 14.68
4,048	$ 14.27	2,151	$ 14.71
3,954	$ 14.36	2,143	$ 14.69
4,197	$ 14.45	1,985	$ 14.70
4,810	$ 14.34	1,810	$ 14.66
5,234	$ 13.97	1,644	$ 14.67
4,383	$ 14.45	1,810	$ 14.69
3,595	$ 14.51	1,721	$ 14.64
3,732	$ 14.37	1,835	$ 14.58
3,276	$ 14.23	1,941	$ 14.61
3,115	$ 14.14	1,865	$ 14.63
2,597	$ 14.14	1,967	$ 14.67
2,436	$ 14.40	2,461	$ 14.62
2,709	$ 14.35	2,160	$ 14.64
2,580	$ 14.38	1,843	$ 14.78
2,724	$ 14.27	1,794	$ 14.76
2,751	$ 13.84	1,904	$ 14.70
2,053	$ 14.12	2,312	$ 14.69

We All Scream for Ice Cream

Consider an ice cream vendor selling ice cream at $2.50 per order with variable costs of $0.75 per order and fixed costs of $3500 per month.

1. Create a breakeven plot for sales volumes between 0 and 3000. What is the breakeven volume?
2. On average, the ice cream vendor sells 2,374 orders per month.
   a. If price went up 10%, keeping everything else constant, what is the absolute change in profits and the percent change in profits?
   b. If sales volume (quantity sold) went up 10%, keeping everything else constant, what is the absolute change in profits and the percent change in profits?
   c. If variable costs went down 10%, keeping everything else constant, what is the absolute change in profits and the percent change in profits?
   d. If fixed costs went down 10%, keeping everything else constant, what is the absolute change in profits and the percent change in profits?
   e. If price went up 10% and volume decreased by 10% at the same time, what is the absolute change in profits and the percent change in profits?
   f. If price went down by 10% and volume increased by 10% at the same time, what is the absolute change in profits and the percent change in profits?
   g. If variable costs went up 10% and volume sold increased by 10% at the same time, what is the absolute change in profits and the percent change in profits?

Chapter 5: Cumulative Distribution Functions and Probability

Distribution tables and plots enable executives and analysts to quickly identify the probability of gaining a specific score from a sample measurement. In many cases, however, the probability of gaining a specific score isn't sufficiently informative. Rather, executives, and therefore analysts, are often more interested in finding the probability of receiving measurements below a specific score (poor performance measurements), measurements above a specific score (high performance measurements), or measurements between two specific scores (as in finding the bulk of the measurements).

To find the probability of receiving a measurement above a score, below a score, or between two scores, analysis need to calculate the cumulative distribution function, often denoted as CDF.

In this chapter, students will learn:

- The meaning of a cumulative distribution function.
- How to find the cumulative distribution function with discrete frequency distributions.
- How to find the cumulative distribution function of a normal distribution given a mean, standard deviation.
- How to use the cumulative distribution function to calculate the probability of finding data above a given score, below a given score, or between any two scores.
- The use of indexes such as the Category Development Index and Brand Development index to describe market territories.

Cumulative Distribution Function

The cumulative distribution function (CDF) describes the probability of obtaining a measurement at or below a given score. It can be described as the function which accumulates the probability of obtaining a measurement up to and including a point, or, more simply, the cumulative probability at a point. We can calculate the cumulative distribution function for many distributions, including those associated with discrete frequency distributions or continuous distributions.

CDF from Discrete Frequency Distributions

We have seen several discrete frequency distributions. In Chapter 2, we examined the frequency distribution of female shoe sizes, a discrete distribution. Discrete distributions are called discrete because they provide information on the probability at given values, but don't provide information regarding the probability at points between two given values.

Returning to our frequency distribution of women's shoe sizes, we had the following distribution function.

Female Shoe Size	Response Frequency
Unknown	6
4	11
5	6
6	75
7	119
8	111
9	66
10	33
11	0
12	1
Total	428

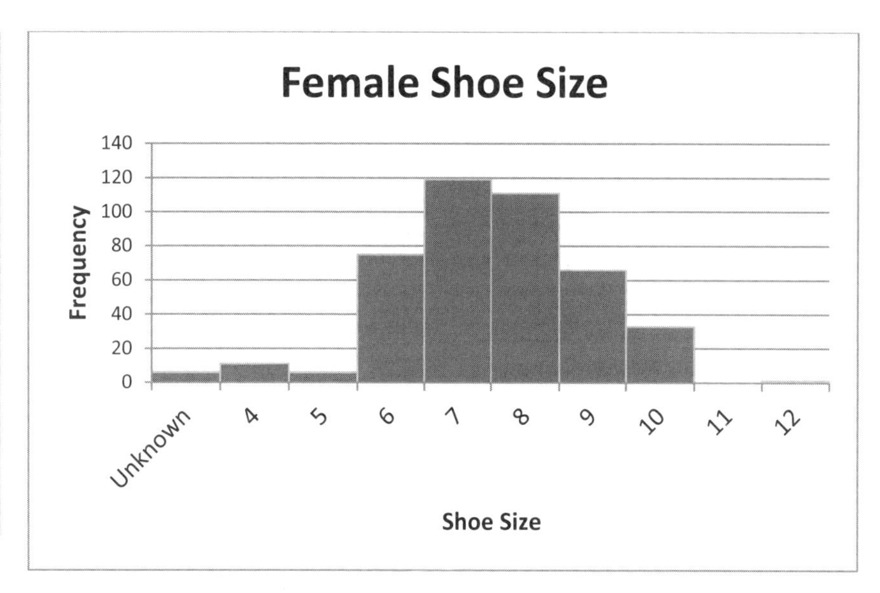

From this frequency response table, we can generate the cumulative distribution function by simply accumulating the probability of gaining a measurement up to and including a point. Again, ignoring the unknown shoe sizes as this data does not provide useful information, we would logically calculate the cumulative distribution function as follows.

- For each shoe size, calculate the probability of a woman having that shoe size.
- Starting from the lowest reported shoe size, in this case a size 4, identify the cumulative probability of having a size 4 or smaller to be that of having a size 4.
- Going to the next higher reported shoe size, a size 5, calculate the cumulative probability of having a size 5 or smaller to be that of having a size 5 plus that of having a size 4.
- Moving to the next higher reported shoe size, a size 6, calculate the cumulative probability of having a size 6 or smaller to be that of having a size 6 plus that of having a size 5 or 4. This calculation can also be stated as: the cumulative distribution at size 6 is equal to the probability of obtaining a size 6 plus the cumulative distribution at size 5.
- We repeat the above calculation until the cumulative distribution for the entire range of sizes has been calculated.

The results would be as shown in the table where the column for the cumulative response probability provides the cumulative distribution function. Notice that the cumulative distribution function ends with 100%. Between the smallest reported size and the highest reported size, a given response has a 100% probability of being provided.

Female Shoe Size	Response Frequency	Response Probability	Cumulative Response Probability
Unknown	6		
4	11	2.6%	2.6%
5	6	1.4%	4.0%
6	75	17.8%	21.8%
7	119	28.2%	50.0%
8	111	26.3%	76.3%
9	66	15.6%	91.9%
10	33	7.8%	99.8%
11	0	0.0%	99.8%
12	1	0.2%	100.0%
Total	428		

It is also useful to plot the cumulative distribution function. For discrete data, it is most common to use a column chart to represent the cumulative distribution function. At times, a line plot will be used.

In Excel, the following formulae were used to generate the above frequency response table.

	A	B	C	D
1	Female Shoe Size	Response Frequency	Response Probability	Cumulative Response Probability
2	Unknown	6		
3	4	11	=B3/SUM(B$3:B$11)	=C3
4	5	6	=B4/SUM(B$3:B$11)	=D3+C4
5	6	75	=B5/SUM(B$3:B$11)	=D4+C5
6	7	119	=B6/SUM(B$3:B$11)	=D5+C6
7	8	111	=B7/SUM(B$3:B$11)	=D6+C7
8	9	66	=B8/SUM(B$3:B$11)	=D7+C8
9	10	33	=B9/SUM(B$3:B$11)	=D8+C9
10	11	0	=B10/SUM(B$3:B$11)	=D9+C10
11	12	1	=B11/SUM(B$3:B$11)	=D10+C11
12	Total	428		

CDF FROM NORMAL DISTRIBUTIONS

In Chapter 4, we introduced our first continuous distribution function, the normal distribution. The normal distribution is called a continuous function because it allows you to find the probability at any point, including any point between any two other points, regardless of how near the two points are.

Calculating the cumulative distribution function from a continuous function requires integral calculus, a skill set beyond that expected of the average marketing analyst. Fortunately, however, normal distributions are both understood and common enough that almost any statistical software will provide a means to calculate the cumulative distribution function of a normal function – even Excel.

If you recall, the Excel function for calculating with the normal distribution, NORMDIST(), has four arguments. The first describes the point under investigation. The second describes the mean score of the distribution. The third describes the standard deviation of the distribution. It is the fourth argument that determines whether the NORMDIST() function returns the normal distribution or the cumulative distribution. The fourth argument, cumulative, requires a true or false input. If true, the NORMDIST() function returns the cumulative distribution at that point. If false, the NORMDIST() function returns the normal distribution at that point.

A plot of the cumulative distribution function of normal distributions with different means and standard deviations reveals many attributes. Using the exact same approach as in Chapter 4 but setting the cumulative argument to "TRUE", we have generated the following examples.

- In the first plot, we are varying the mean score while holding the standard deviation in scores constant. As the mean goes from a small number to a larger number, the cumulative distribution function goes from 0 and 1 at different measurements.

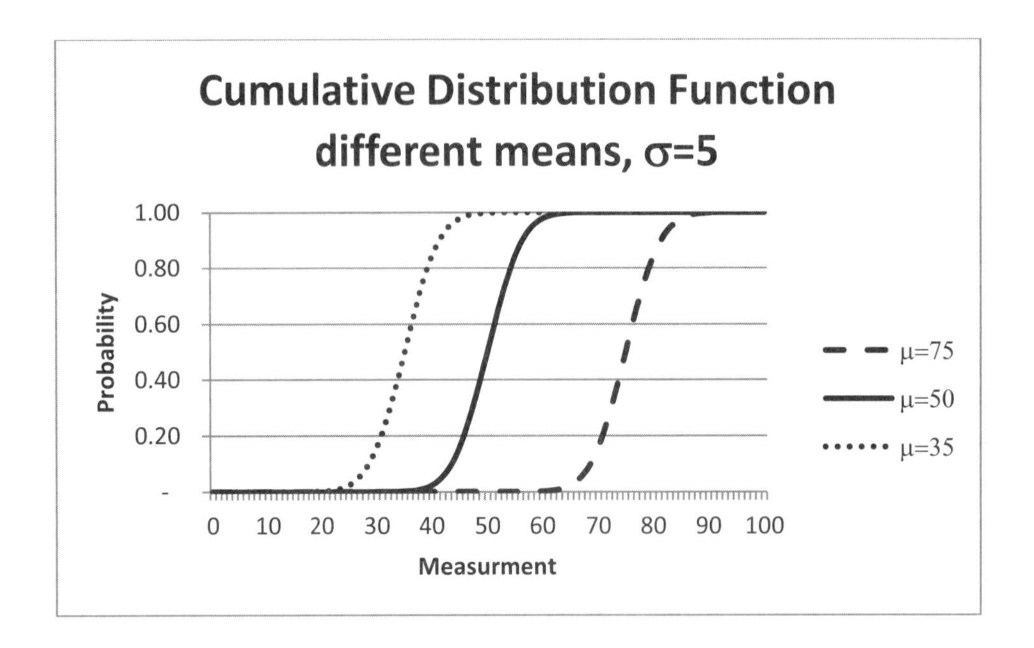

- In the second plot, we are varying the standard deviation while holding the mean constant. As the standard deviation goes from a small number to a larger number, the cumulative distribution function goes from sharply changing between 0 and 1 to smoothly increasing from 0 to 1.

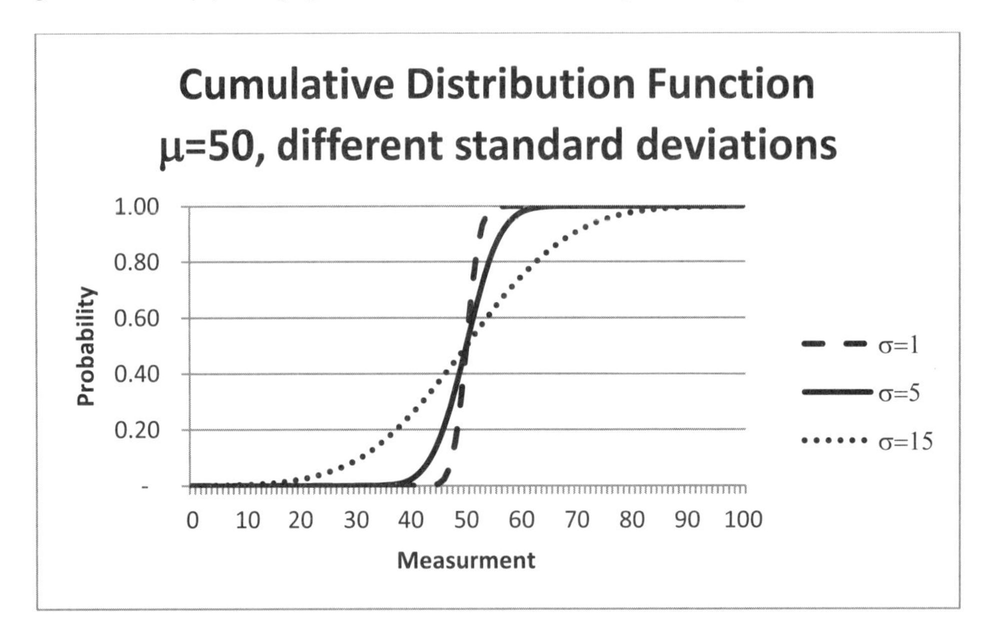

Importantly for marketing, new product launches often follow the cumulative distribution function of the normal distribution. Initially, sales are slow. Over time, sales begin to pick up. At some point in time, sales penetrate 50% of the potential market. At the 50% market penetration point, the growth curve, like the cumulative distribution function, strikes the inflection point, at which further growth comes slower. Past the inflection point, the growth slows until the market enters maturity.

PROBABILITY AND CUMULATIVE DISTRIBUTION FUNCTIONS

The cumulative distribution function is useful for finding the probability between any two scores. While this is important for some discrete distribution, it is imperatively useful for continuous functions. Rarely does a researcher desire the probability of obtaining a score of exactly 6.000... on a scale of 1 to 10 in a continuous distribution. Rather, they might be interested in the probability of obtaining a score of approximately 6, say above 5.5 but below 6.5. To find this probability, we must use the cumulative distribution function.

PROBABILITY OF OBTAINING MEASUREMENTS AT OR BELOW A GIVEN SCORE

As stated, the cumulative distribution function provides the probability of obtaining a measurement at or below a given score.

Thus, for a discrete function, the probability of obtaining a measurement at or below a given score is given by the cumulative distribution function as described above. The analyst will calculate the probability of obtaining each score, then calculate the cumulative probability up to and including that score, and finally read the probability of obtaining a measurement at or below a given score from the table that describes the cumulative distribution function.

For a normal distribution function, the probability of obtaining a measurement at or below a given score is given by the cumulative distribution function as described above. With Excel, the analyst will simply use the NORMDIST() function with the distribution's mean, standard deviation and set the fourth argument to "TRUE" to attain the cumulative distribution at the point under investigation.

PROBABILITY OF OBTAINING MEASUREMENTS AT OR ABOVE A GIVEN SCORE

To find the probability of obtaining a measurement at or above a given point, we can deploy one of two methods depending upon the type of distribution given. For this and the following section, we will begin with a discussion of continuous distributions before turning to discrete functions due to a small additional complexity associated with discrete functions.

For continuous functions, the probability of finding a measurement at or above a given point is simply 1 minus the cumulative distribution function. This property arises logically from the fact that, across all possible responses, the probability of obtaining some response is 100%, or 1, and that the cumulative distribution function is precisely the probability of finding a measurement at or below a given point. Thus, if the probability of finding a measurement at or below a given point is given by the CDF at that point, and the probability of finding a measurement anywhere is 1, then the probability of finding a measurement at or above a given point is given by 1-CDF at that point.

Using Excel, we would find the probability at or above a given point with the expression of "=1-NORMDIST(x,μ,σ,TRUE)".

For a discrete function, the probability of obtaining a measurement at or above a given score requires only a slight modification from the above discussion. Recall that the cumulative distribution function includes the probability of obtaining a measurement at a point. 1-CDF of a discrete distribution would yield the probability of obtaining a measurement at a score above the current score, and would not include the probability of obtaining the current score. Hence, one approach of calculating the probability of obtaining a measurement at or above a given score is to calculate the value of 1 less than the cumulative distribution at that point, and then add back in the probability at that point. For a continuous distribution, this correction is unnecessary. For discrete distributions, this correction ensures accuracy and clarity in the calculation.

An alternative approach for finding the probability of obtaining a measurement at or above a given point that is very similar to that described in calculating the cumulative distribution function requires repeating the process used for calculating the cumulative distribution function but doing so backwards. In calculating the cumulative distribution function for discrete data, we started at the lowest measurement and accumulated probability up to the highest measurement. Thus, to find the probability of obtaining a measurement at or above a given point, we can repeat the procedure, but this time starting with the highest measurement and accumulating probability down to the lowest measurement.

Continuing with our women's shoes example, we would calculate the following:

Female Shoe Size	Response Frequency	Response Probability	Cumulative Response Probability	Probability of obtaining a shoe size at or above that given
Unknown	6			
4	11	2.6%	2.6%	100.0%
5	6	1.4%	4.0%	97.4%
6	75	17.8%	21.8%	96.0%
7	119	28.2%	50.0%	78.2%
8	111	26.3%	76.3%	50.0%
9	66	15.6%	91.9%	23.7%
10	33	7.8%	99.8%	8.1%
11	0	0.0%	99.8%	0.2%
12	1	0.2%	100.0%	0.2%
Total	428			

Where the following formulae for the final column was used:

	A	B	C	D	E
1	Female Shoe Size	Response Frequency	Response Probability	Cumulative Response Probability	Probability of obtaining a shoe size at or above that given
2	Unknown	6			
3	4	11	2.6%	2.6%	=E4+C3
4	5	6	1.4%	4.0%	=E5+C4
5	6	75	17.8%	21.8%	=E6+C5
6	7	119	28.2%	50.0%	=E7+C6
7	8	111	26.3%	76.3%	=E8+C7
8	9	66	15.6%	91.9%	=E9+C8
9	10	33	7.8%	99.8%	=E10+C9
10	11	0	0.0%	99.8%	=E11+C10
11	12	1	0.2%	100.0%	=C11
12	Total	428			

Probability of Obtaining Measurements between Two Scores

When we are calculating the probability of obtaining a measurement between two scores, we are searching for the probability of obtaining a measure in a range of scores. That range of scores has an upper limit and a lower limit.

For continuous functions, the probability of finding a measurement between the upper limit and lower limit is simply the difference of the cumulative distribution at the upper limit the cumulative distribution at the lower limit. This property arises logically from the fact that the cumulative distribution at the upper limit includes the cumulative distribution at the lower limit. All we are doing is subtracting off the unnecessary probability from that obtained at the upper limit to find that desired of the range.

Thus, in Excel, we would find the probability of obtaining a measurement between two points with the expression of =NORMDIST(UL,μ,σ,TRUE) - NORMDIST(LL,μ,σ,TRUE), where we have denoted the upper limit as UL and the lower limit as LL.

For discrete functions, the most straightforward means of calculating the probability of obtaining a score within a range is to add the probabilities of obtaining a measure at any specific score between the upper and lower limits of the range of interest with the endpoints usually included.

Marketing Metrics: Indexing

Indexes can be used in many situations. To introduce the use of indexes in marketing decisions, let us consider CDI and BDI and then provide an example of indexing in business markets.

Category Development Index

The Category Development Index (CDI) describes the level of sales development of a specific category in a Metropolitan Statistical Area, in comparison to the sales of the general commodity group which contains that specific category in that Metropolitan Statistical Area. In consumer markets, a Metropolitan Statistical Area is commonly referred to as a market, and sales volumes are commonly measured in dollars and simply referred to as volume.

To calculate the CDI for a consumer product:

- Calculate the **Category Sales Percentage** as the fraction of revenue earned from a specific market to that of the entire nation for a specific category. (Revenue is casually referred to as "volume").
- Calculate the **All Commodity Sales Percentage** as the fraction of revenue earned from a specific market to that of the entire nation for the general commodity group in which that specific category resides.
- Calculate the Category Development Index in a market as the ratio of the Category Percentage Sales to the All Commodity Sales Percentage, times 100.

For example, a marketing manager is interested in the performance of pickle sales in different metropolitan statistical areas and holds the following data from NPD sales records and US Census information. Pickles are a specific product category within the more general commodity group of condiments. The national market for pickles was $1.095 billion, and the national market for condiments was $39.05 billion. Sales for specific markets are shown below.

	All Condiment Commodity Volume	Pickle Category Volume
	ACV	CV
Nationally	$ 39,050,000,000	$ 1,095,000,000
Atlanta	$ 550,605,000	$ 20,889,000
Boston	$ 456,885,000	$ 10,101,000
Los Angeles	$ 1,276,935,000	$ 36,894,000

Given this data, the analyst will calculate the Pickle Category Sales Percentage, All Condiment Commodity Sales Percentage, and Category Development Index for each market.

	All Condiment Commodity Volume	Pickle Category Volume	Commodity Sales Percentage	Category Sales Percentage	Category Development Index
	ACV	CV	ACSP	CSP	CDI
National	$ 39,050,000,000	$ 1,095,000,000			
Atlanta	$ 550,605,000	$ 20,889,000	1.41%	1.91%	135
Boston	$ 456,885,000	$ 10,101,000	1.17%	0.92%	79
Los Angeles	$ 1,276,935,000	$ 36,894,000	3.27%	3.37%	103

Where, for Atlanta, we calculate the All Commodity Sales Percentage as

$$ACSP = \frac{Atlanta\ Condiment\ All\ Commodity\ Volume}{National\ Condiment\ All\ Commodity\ Volume} = \frac{\$\quad 550{,}605{,}000}{\$\ 39{,}050{,}000{,}000} = 1.41\%$$

Furthermore, we calculate the Category Sales Percentage as

$$CSP = \frac{Atlanta\ Pickle\ Category\ Volume}{National\ Pickle\ Category\ Volume} = \frac{\$\quad 20{,}889{,}000}{\$\ 1{,}095{,}000{,}000} = 1.91\%$$

And the Category Development Index as

$$CDI = \frac{Atlanta\ CSP}{Atlanta\ ACSP} \cdot 100 = \frac{1.91\%}{1.41\%} \cdot 100 = 135$$

From these calculations, the analyst reports that the Atlanta market constitutes 1.41% of all condiment sales nationally, 1.91% of all pickle sales nationally, and has a category development index of 135 – a rather well developed market from a category perspective.

A high or low CDI does not imply that more or less marketing is required, or that more or fewer products are needed. CDI merely indicates the level of concentration of customers who purchase from the specific category of interest.

In general, market researchers seeking to know more about a specific category should concentrate their research on markets with a high CDI and avoid markets with a low CDI, because markets with a high CDI are likely to be informed of the product category. Because the Atlanta market is well developed, it represents a good market for doing market research on pickles in general. In contrast, the Boston market is relatively poorly developed at the category level and, as such, would be a more questionable market for conducting market research on pickles.

Brand Development Index

The Brand Development Index (BDI) describes the level of development of market brand sales in comparison to category sales.

To calculate the BDI for a consumer product:

- Calculate the **Brand Sales Percentage** as the fraction of revenue derived from of a specific brand in a market to that of the nation. (Revenue associated with a brand is casually referred to as "brand volume").
- Calculate the Brand Development Index in a market as the ratio of the Brand Sales Percentage to the Category Sales Percentage, times 100.

For example, the marketing manager is not simply interested in the pickle market as a whole, but perhaps for Vlassic pickles in particular, and holds the following data.

	Pickle Category Volume	Vlassic Volume
	CV	BV
Nationally	$ 1,095,000,000	$ 386,000,000
Atlanta	$ 20,889,000	$ 6,818,335
Boston	$ 10,101,000	$ 1,577,583
Los Angeles	$ 36,894,000	$ 9,722,810

Given this data, the analyst will calculate the Brand Sales Percentage, Category Sales Percentage, and Brand Development Index for each market.

	Pickle Category Volume	Vlassic Volume	Category Sales Percentage	Brand Sales Percentage	Brand Development Index
	CV	BV	CSP	BSP	BDI
Nationally	$ 1,095,000,000	$ 386,000,000			
Atlanta	$ 20,889,000	$ 6,818,000	1.91%	1.77%	93
Boston	$ 10,101,000	$ 1,577,000	0.92%	0.41%	44
Los Angeles	$ 36,894,000	$ 9,722,000	3.37%	2.52%	75

Where, for Atlanta, we calculate the Brand Sales Percentage as

$$BSP = \frac{Atlanta\ Vlassic\ Volume}{National\ Vlassic\ Volume} = \frac{\$\ \ \ 6{,}818{,}000}{\$\ 386{,}000{,}000} = 1.77\%$$

And the Brand Development Index as

$$BDI = \frac{Atlanta\ BSP}{Atlanta\ CSP} \cdot 100 = \frac{1.77\%}{1.91\%} \cdot 100 = 93$$

Atlanta's Category Sales Percentage was calculated above.

From these calculations, the analyst reports that the Atlanta market constitutes 1.77% of all Vlassic sales nationally and has a brand development index of 93 – slightly below Vlassic's better developed markets.

A high or low BDI does not imply that more or less marketing is required, or that more or fewer products are needed. Marketer's proclivity is to pour more resources into low BDI markets. However, there may be barriers to market development within specific markets. For instance, Vlassic competes with a number of Polish brands of pickles in Chicago which would be difficult to displace due to cultural preferences.

BDI primarily indicates the level of concentration of customers who purchase the specific brand of interest. In other words, BDI indicate whether the percentage of customers purchasing the focal brand in a given market is similar to what would be expected based on category sales for that market.

In general, market researchers seeking to know more about a specific brand should concentrate their research on markets with a high BDI and avoid markets with a low BDI, because markets with a high BDI are more likely to be knowledgeable about the brand. Because the Atlanta market is somewhat well developed, it represents a good market for doing market research on Vlassic. In contrast, Boston is very poorly developed at the brand level and, as such, would be a very questionable market for conducting market research on Vlassic.

Business Markets

Index numbers can be useful for identifying the sales penetration within a specific business customer. For instance, suppose an office supplier discovers that, on average, spending on office supplies is $32.73 per employee. For three different customers, the office supplier indexes sales against expectations, given his knowledge of the size of the companies, by taking the ratio of the sales obtained to the sales expected, times 100.

Firm	Employees	Sales	Obtained Sales / Employee	Expected Sales / Employee	Penetration Index
A	97	$ 1,429	$ 14.73	$ 32.73	45
B	32	$ 1,435	$ 44.84	$ 32.73	137
C	750	$ 6,137	$ 8.18	$ 32.73	25

While this office supplier is doing well with firm B, she is missing a large portion of the potential revenue from firm C, which is also a larger company. From this data, the office supplier may determine it to be in her best interest to increase the sales and account management effort for firm C.

Exercises

Kale

A survey was conducted asking for the level of agreement among a sample. The following data was collected and tabulated.

Response Items	Strongly Disagree	Disagree	Neither Agree nor Disagree	Agree	Strongly Agree
Encoding	1	2	3	4	5
I like Kale.	2	7	5	4	3

1. Calculate the probability of obtaining a specific response in a measurement.
2. Calculate the cumulative probability function of response.
3. What fraction of people like kale at a score of 3 or lower?
4. What fraction of people like kale at a score of 4 or higher (also known as a "Top 2 Box" score)?
5. Plot the cumulative distribution function of responses with a column chart.

Men's Shoes

A sample of males was surveyed regarding their shoe sizes and data was collected. Ignore uninformative data points in this exercise.

1. Calculate the probability of obtaining a specific shoe size in a given measurement.
2. Calculate the cumulative probability function of men's shoe sizes.
    a. What percentage of men wears a size 8 or larger?
    b. What percentage of men wears a size 10 or larger?
    c. What percentage of men wears a size 12 or larger?
3. Plot the cumulative distribution function of men's shoe sizes with a bar chart.
4. Calculate the probability of having a shoe size equal to or higher than a given shoe size.
    a. What percentage of men wears a size 8 or smaller?
    b. What percentage of men wears a size 10 or smaller?
    c. What percentage of men wears a size 12 or smaller?

Male Shoe Size	Response Frequency
Unknown	6
6	17
7	2
8	51
9	154
10	172
11	94
12	51
13	21
14	5

WEIGHTY CHALLENGE FOR MALES

In 2002, the average weight of an adult American male was 191 lbs. Assume that the weights of American adult males are normally distributed with a standard deviation of 22 lbs.

1. Plot a normal distribution curve for the weight of a male. Consider weights as low as 120 lbs and as high as 350 lbs.
2. Plot a cumulative distribution function of the distribution of the weights of males. Consider weights as low as 120 lbs and as high as 350 lbs.
3. What percentage of males weighs less than 150 lbs?
4. What percentage of males weighs less than 175 lbs?
5. What percentage of males weighs more than 175 lbs?
6. What percentage of males weighs more than 191 lbs?
7. What percentage of males weighs more than 220 lbs?
8. What percentage of males weighs more than 250 lbs?
9. What percentage of males weighs more than 300 lbs?
10. What percentage of males weighs between 150 and 175 lbs?
11. What percentage of males weighs between 175and 215 lbs?
12. What is the range of weights that correspond to 1 standard deviation? What percentage of the male population weighs within this range?
13. What is the range of weights that correspond to 1.96 standard deviations? What percentage of the male population weighs within this range?
14. What is the range of weights that correspond to 2.58 standard deviations? What percentage of the male population weighs within this range?

A Challenge for Females

Assume that the weights of American adult females are normally distributed.

1. Plot a normal distribution curve for the weight of a female using Z scores on the X axis between -3 and + 3 standard deviations.
2. Plot a cumulative distribution function of the distribution of the weights of females between -3 and +3 standard deviations.
3. What percentage of females weighs 2.58 standard deviations below the average?
4. What percentage of females weighs 1.96 standard deviations below the average?
5. What percentage of females weighs 1.00 standard deviations below the average?
6. What percentage of females weighs 1.00 standard deviations above the average?
7. What percentage of females weighs 1.96 standard deviations above the average?
8. What percentage of females weighs 2.58 standard deviations above the average?
9. What percentage of females weighs between -1 and +1 standard deviations from the average?
10. What percentage of females weighs between the mean weight and +1 standard deviations above the average?
11. What percentage of females weighs between -1.96 and +1.96 standard deviations from the average?
12. What percentage of females weighs outside of the range of -1.96 and +1.96 standard deviations from the average?
13. What percentage of females weighs between -2.58 and +2.58 standard deviations from the average?
14. What percentage of females weighs outside of the range of -2.58 and +2.58 standard deviations from the average?

Firm Performance

Sales performance data over 26 weeks for a firm were collected.

1. Create a probability distribution plot of the data using 7 intervals.
2. On a separate chart, plot the cumulative distribution function of this data.
3. What is the median and standard deviation in units sold?
4. Find the expected percentage of sales in a given week if the sales followed a normal distribution by taking the difference of the normal cumulative distribution at the upper and lower limit of each interval used in calculating the probability distribution. Use the median and standard deviation calculated above.
5. Plot the expected percentage of sales in a given week if the sales followed a normal distribution as a column chart on the same plot as the discrete probability distribution. Select the column chart series related to the normal distribution and change its, and only its, plot type to a line.
6. Do the sales data follow a normal curve? How would you explain the difference, if any?

Week	Unit Sales	Week	Unit Sales
1	10,662	14	14,180
2	15,119	15	15,398
3	16,751	16	19,882
4	13,493	17	15,951
5	9,573	18	14,732
6	14,284	19	28,056
7	15,244	20	15,375
8	22,138	21	16,843
9	34,342	22	12,795
10	18,446	23	13,131
11	11,544	24	11,004
12	11,940	25	14,206
13	10,162	26	10,440

GRADUATES VS. NON-GRADUATES PURCHASING

A large bookstore chain wanted to target its customers. They decided to see whether purchases of college graduates differed significantly from those of non-college graduates. The number of books purchased yearly by each group is shown below. Assume homogeneity of variances. Samples were randomly selected from the store's data base.

1. Create a probability distribution plot of the data using the 5 intervals which span both datasets.
2. On a separate chart, plot the cumulative distribution function for both datasets.
3. What is the mean and standard deviation for each dataset in books purchased?
4. Find the expected percentage of books purchased per year.
5. Compare the observed sales to the sales that would have been predicted if the sales followed a normal curve.
    a. Identify the expected sales within an interval by taking the difference of the normal cumulative distribution at the upper and lower limit of each interval used in calculating the discrete probability distribution.
    b. Use the sample means and standard deviations for calculating the normal distributions.
6. Plot the expected percentage of book purchases if purchases followed a normal distribution as a line plot over the percentage probability distribution plot of the data.
7. Do the book purchases follow a normal curve? How would you explain the difference, if any?

NUMBER OF BOOKS PURCHASED IN PAST YEAR	
College Grads	Non-College Grads
12	10
20	21
18	12
16	9
22	15
15	16
17	14
21	11
19	16
25	7
20	17
16	8

Vlassic in Chicago and San Antonio

The following sales data has been collected regarding Vlassic Pickles:

	All Condiment Commodity Volume	Pickle Category Volume	Vlassic Volume
	ACV	CV	BV
Nationally	$ 39,050,000,000	$ 1,095,000,000	$ 386,000,000
Chicago	$ 893,000,000	$ 24,327,000	$ 2,423,000
San Antonio	$ 487,000,000	$ 11,982,000	$ 5,452,000

1. For Chicago and San Antonio:
    a. Calculate the All Commodity Sales Percentage.
    b. Calculate the Category Sales Percentage.
    c. Calculate the Brand Sales Percentage.
    d. Calculate the Category Development Index.
    e. Calculate the Brand Development Index.
2. In which of these markets would you suggest research be conducted on pickles?
3. In which of these markets would you suggest research be conducted on Vlassic?

Chocolate Index

30% of Hershey's sales are in Chicago, 8% are in St. Louis and 3% are in Atlanta. Atlanta accounts for 1% of cocoa sales, whereas St. Louis has 4% and Chicago has 15%.

1. Calculate the BDI's for each of these three markets.
2. Explain what the index number means for each market to Hershey's cocoa marketers.

MARKETING PRINTING INDEX

A specialty printing company creates banners for tradeshows. On average, a corporation spends $1830 on trade show printing per year for every million dollars of revenue. The specialty printer wants to index his penetration into three accounts.

Firm	Revenue (millions)	Trade Show Printing Sales	Obtained Sales / Revenue (M)	Expected Sales / Revenue (M)	Penetration Index
A	20	$ 12,328		$ 1830	
B	8	$ 16,983		$ 1830	
C	173	$ 274,328		$ 1830	

1. Calculate the obtained sales per million in revenue for each firm (M means million).
2. Calculate the penetration index for each firm.
3. In which firm is the specialty printing company doing best?
4. In which firm is the specialty printing company doing worst?

CHAPTER 6: STUDENT'S T-TEST FOR NUMERICAL MEASUREMENTS

For the first five chapters, we have been examining approaches for investigating and describing frequency distributions. We have seen how measurements can yield a variety of responses, how responses can be scored, and how scores can be characterized by their central tendency and distribution. Furthermore, we have seen how the distribution of scores can be described in terms of the spread and shape. We have also examined ways of calculating the probability of receiving a specific score, a score equal or higher than a minimum cutoff score, equal or less than a maximum cutoff score, or a score between any two limits.

However, we have not yet been able to state whether two distributions are definitively similar or different, or whether two distributions are related or unrelated. In this and the remaining four chapters, we will introduce the challenge of characterizing the relationship between two sample measurements. We begin with an approach for examining the similarity of numerical measurements across samples: Student's t-test.

In this chapter, students will learn:

- The challenge of determining whether two samples came from two different populations or from the same population.
- How hypothesis testing can be used to guide decisions regarding the similarity or differences between sample measurements.
- How to use Student's t-tests to describe the likelihood that two different samples investigated with the same numerically scored measurements arise from different populations or the same population.
- How to use ratios to estimate market size and demand for a product or service.

POPULATIONS, SAMPLES, AND DIFFERENCES

A measurement of an item on a sample from a population will often yield a distribution of scores. If we repeat the same measurement on a different sample from the same population, it is likely we will receive a slightly different but relatively similar distribution of scores. Even though there may be differences in the distributions between the two measurements, these differences could be attributed to random sample error since both samples are from the same population.

In contrast, if we measure an item from samples from two different populations, we would not hold the same expectation regarding the outcome of the measurements. In fact, if the populations differ along a dimension of interest, and that dimension of interest is being measured, we would expect the measurements to reveal differences. The differences might arise in the central tendency of the distribution of scores, or even in the spread and shape of the distribution of scores. And, because the samples are from two different populations, we would consider these differences as being a reflection of true differences between populations.

Yet this leads to a challenge regarding the characterization of differences between measures between two samples. How should a researcher determine if the measured differences arise from (1) random sample error in measurements on the same population or (2) true differences between two different populations being reflected in different measurements?

For example, suppose a researcher selling softened toilet paper is interested to know if the gender of the customer determines the likelihood of purchasing this type of product. The researcher then collects a sample, named for research purposes Sample A, consisting of individuals who identify themselves as male. The researcher also collects a second sample, named for research purposes Sample B, consisting of individuals who identify themselves as male. After measuring both samples according to likelihood of purchase, it is highly likely that a slight difference will be found, but this difference is insignificant. In other words, both Sample A and Sample B are equally likely to purchase softened toilet paper. Hence, the researcher would conclude that, with respect to the specific issue of softened toilet paper, individuals from different genders are from the same population group and any differences can be attributed to random sampling errors.

Next, the researcher is interested to know if political party affiliation determines the likelihood of purchasing from the category of softened toilet paper. The researcher then collects a sample, named for research purposes Sample C, consisting of individuals who identify themselves as supporting the Green Party. The researcher also collects another sample, named for research purposes Sample D, consisting of individuals who identify themselves as supporting the Democratic or Republican Party. After measuring both samples according to likelihood of purchase, it is highly likely that a real difference will be found, and that this difference is significant. Sample C and Sample D are relatively different in terms of their likelihood of purchasing softened toilet paper. Hence, the researcher would conclude that, with respect to the specific issue of softened toilet paper, individuals that support different political parties are from different population groups, and any measured differences reflect a true underlying difference between the population groups.

To determine that the difference in measurements between Sample A and B were due to random sampling error while the difference in measurements between Sample C and D were due to actual differences between the populations, the analyst will conduct a statistical analysis and interpret the meaning of an analysis statistic using hypothesis testing.

Hypothesis Testing

The hypothesis test is an approach for testing the validity of an assumption.

For quantitative research purposes, the starting assumption is that everything is the same. We could state this as:

- The same measurement applied to two different samples will yield the same result.
- Any difference in measurements between two samples is due to random fluctuations and is therefore insignificant and meaningless.
- Differences in measurements between two samples always arise from random sampling errors.

This assumption is often true, for people are alike in more respects than not (we all must eat and sleep, we all tend to benefit from meaningful relationships, and so on). Mostly true, however, isn't always true. Every person is unique and different in at least some dimension.

Hence, this starting assumption is sometimes false. Or, more fully stated, the same measurement applied to two different samples may yield different results, and that difference in measurements between two samples may be significant. That is, difference in measurements between two samples may be due to a real difference between two different population groups.

In a hypothesis test, we call the starting assumption the Null Hypothesis. The alternative to the Null Hypothesis is the Alternative Hypothesis. Hence, for a statistical test of differences between sample measurements:

- **The Null Hypothesis assumes that the difference between sample measurements is insignificant and arises from random sampling error.**
- **The Alternative Hypothesis includes all situations where the Null Hypothesis is false, that is that the difference between sample measurements is significant and arises from real differences between the population groups.**

In every statistical analysis, either the Null Hypothesis is true and Alternative Hypothesis is false, or the Null Hypothesis is false and Alternative Hypothesis is true. They can't both be false at the same time, and they can't both be true at the same time.

To conduct a hypothesis test, we perform the following logic steps:

1. Assume the Null Hypothesis is True.
2. Attempt to prove the Null Hypothesis False.
    a. If we cannot determine that the Null Hypothesis is False, we accept the Null Hypothesis.
    b. If we can determine that the Null Hypothesis is False, we reject the Null Hypothesis and accept the Alternative Hypothesis.

Hence, in a statistical analysis of the differences between two sample measurements, we

1. Assume that the two sample measurements are relatively the same.
2. We test this assumption using a statistical test of significance.
    a. If we cannot statistically determine that the differences are significant, we accept the assumption that the two sample measures are relatively the same and any differences arises from random sampling error.
    b. If we can statistically determine that the differences are significant, we reject the assumption that the differences between the two sample measures are relatively the same, and accept that the two sample measurements are different due to differences between the two population groups.

Statistical Test of Significance

The hypothesis testing procedure described above requires a statistical test of significance. One type of statistical test relies on a calculation of the probability that the two sample measurement distributions differ due to random sampling errors alone, and then comparing that calculated probability with a predetermined cutoff value for the test of significance.

p-Value

The p-value is a statistic used to measure the probability that the two sample measurement distributions differ due to random sampling errors alone. p-Values range from 0 to 1. If the p-value statistic is 1, then there was no difference detected between the two sample measurement distributions. As the p-value decreases from 1 to zero, the detected differences between the two sample measurement distributions increase.

In general, p-values near 1 indicate that sample measurement distributions are relatively similar; therefore we accept the assumption that the two samples are relatively similar. In contrast, p-values near 0 indicate that sample measurements are relatively different; therefore we accept the assumption that the two samples are relatively different.

But how near to zero must a p-value be before we determine that the samples are relatively different? That is, at what level do we determine the probability that the two sample measurement distributions differ due to random sampling errors alone to be so small that we should reject this null hypothesis? We need some sort of cutoff to use to determine if the p-value is close enough to zero for us to feel comfortable rejecting the null hypothesis and accepting the alternative hypothesis.

ALPHA

The cutoff we use is called the significance level, often denoted by alpha (the Greek letter α). We select significance level on the basis of the level of reliability we demand from our statistical test of significance.

In a statistical test of significance, we compare alpha to the p-value.

- If the p-value is greater than the chosen significance level, we claim
    - That the probability that the two sample measurement distributions differ due to random sampling errors alone is sufficiently large that is...
    - We accept the null hypothesis, that is...
    - That the differences between the two sample measurement distributions are due to random sampling errors, that is...
    - The differences are statistically insignificant.
- If the p-value is equal to or less than the chosen significance level, we claim
    - That the probability that the two sample measurement distributions differ due to random sampling errors alone to be extremely small, that is...
    - We reject the null hypothesis and accept the alternative hypothesis, that is...
    - That the differences between the two sample measurement distributions are due to real differences between the population groups, that is...
    - The differences are statistically significant.

Clearly, the chosen significance level is very important. What significance level should we choose? How sure do we want to be before we declare things to be different or the same?

If we choose the significance level to be 5% ($\alpha = 0.05$), then we are choosing our cutoff such that the likelihood that the differences in measurements with a p-value less than alpha are due to random sample error, rather than being a true difference, is less than 1 in 20. For most work in marketing, this level of confidence is sufficient for determining significance.

At times, we may choose other significance levels. When we do, we are changing the requirement for determining whether differences are real or due to sampling error.

If alpha equals ...	Then the significance level is	And the likelihood that the same population group would yield sample measurements with a p-value less than the significance level is
.25	25%	1 in 4
.10	10%	1 in 10
.05	5%	1 in 20
.01	1%	1 in 100

STUDENT'S T-TEST (STUDENT IS THE LAST NAME OF MATHEMATICIAN)

Student's t-test is used for comparing two different sample distributions of numerical data where the samples were subjected to the same measurement. A student's t-test reveals the probability that two sample measurement distributions differ due to random sampling errors alone.

The t-test relies upon a t-distribution.

The t-distribution describes the expectation of obtaining a score when using a sample rather than the full population. With only a set of sample measurements, the analysts will not know the population's standard deviation, only the sample's standard deviation. As such, the t-distribution is a more appropriate distribution for defining the range of expected scores. When the population standard deviation is known, analysts may use the normal distribution. In other situations, the t-distribution is more appropriate when comparing samples.

In shape, the t-distribution is very similar to the normal distribution, but a little broader for small sample sizes.

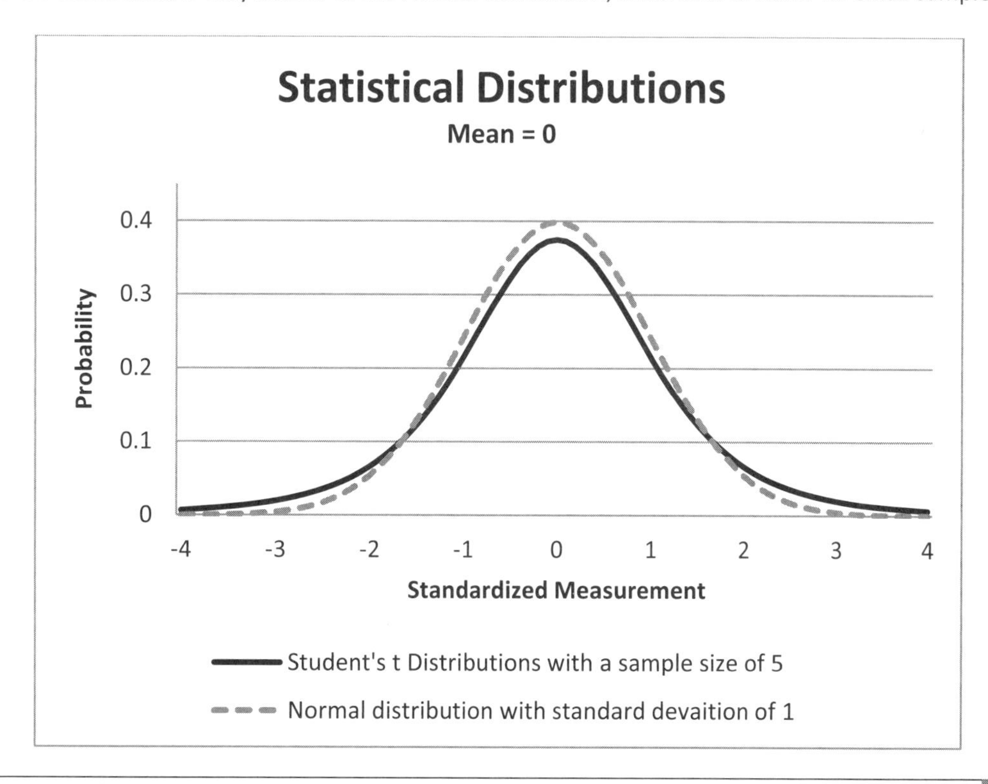

When conducting a t-test, the analyst uses the t-distribution to identify the probability that the two sample measurements differ due to random sampling error alone. To identify this probability, there are two decisions that must be made: (1) Did the researcher expect one sample to yield a definitively lower score than the other, and (2) what type of sampling was done? The first decision determines the number of tails to be used in the t-test, while the second decision determines the specific type of t-test to deploy.

TAILS

The number of tails used in a t-test is determined by the expectations held by the researcher prior to conducting the research. If the researcher expects one of the samples to yield a higher measure than the other, the analyst will conduct a one-tailed t-test. Otherwise, the researcher will conduct a two-tailed t-test.

In a two-tailed t-test, the researcher does not expect that the one sample average should be greater than the other. For instance, the researcher may hold no suspicion that people who self-identify as male (Sample A) are likely to demand softened toilet paper more or less than people who self-identify as female (Sample B). If so, the researcher is attempting to detect any difference between the samples. Either Sample A contains a higher set of scores than Sample B, or Sample B contains a higher set of scores than Sample A. But, a priori, the researcher does not hold an expectation with respect to which sample has a higher average score.

In this situation, the researcher is looking for measurements at either tail of the distribution of expected measurements. Therefore, even prior to calculating the sample means, the researcher will decide to conduct a two-tailed t-test.

If we look at the Student's distribution about a mean, the two-tailed t-test examines the likelihood that the difference in means falls above or below a given score at the opposite end of the distribution. In other words, it looks at both tails of the student's distribution.

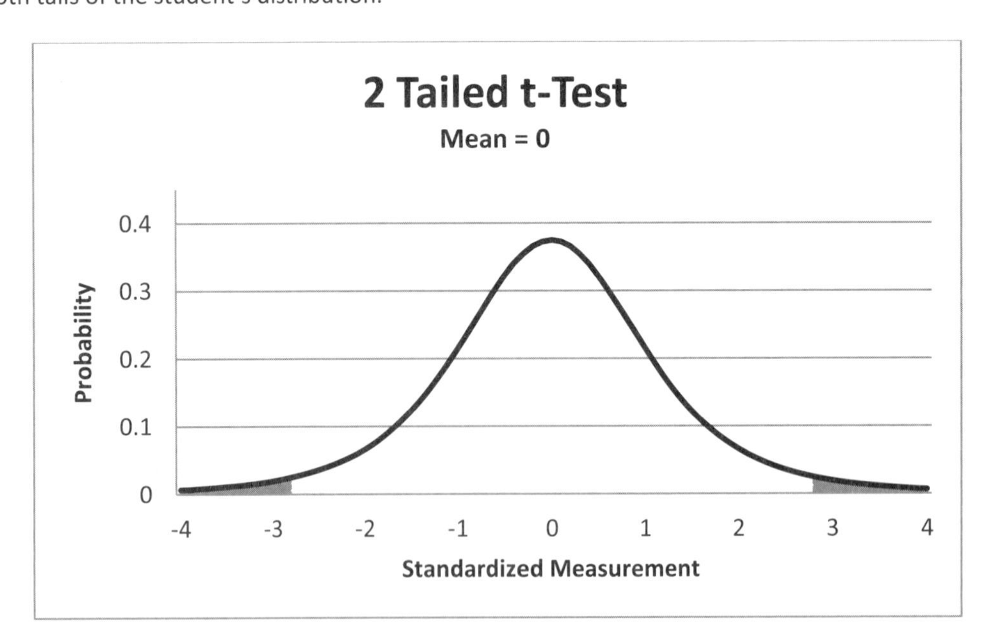

In contrast, in a one-tailed t-test, the researcher expects that the one sample average should be greater than the other. For instance, the researcher may suspect that individuals whose political affiliation tends to be Democrat or Republican (Sample D) exhibit a higher likelihood of purchasing softened toilet paper than people whose political affiliation tends to be Green Party (Sample C). If this is the case, the researcher is attempting to demonstrate that Sample D contains a higher set of scores than sample C.

In this situation, the researcher is looking for measurements that fall at the upper end of the distribution of expected measurements. Therefore, even prior to calculating the sample means, the researcher will determine to do a one-tailed t-test.

If we look at the Student's distribution about a mean, the one-tailed t-test examines the likelihood that the difference in means falls above a given score. In other words, it looks at just one tail of the student's distribution.

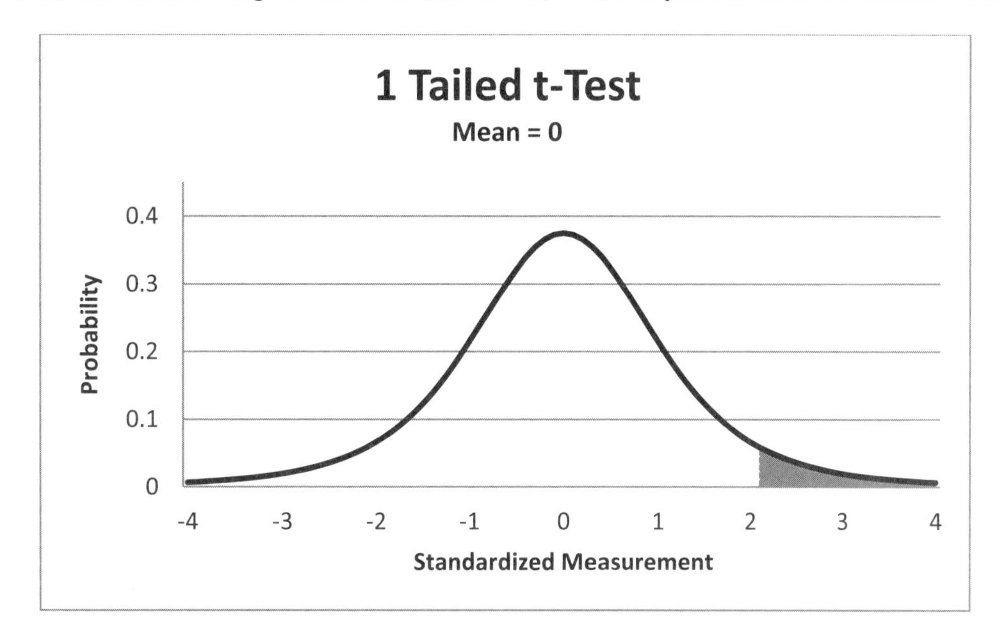

The choice of using a one-tailed or two-tailed t-test is a decision that must be made by the analyst based on the researcher's expectations. If the research expects, and is seeking to demonstrate, that a specific sample has a higher average than the others, the analyst will conduct a one-tailed t-test. In all other cases, the analyst will conduct a two-tailed t-test. That is, in all situations in which the research has not led the analyst to expect a specific result, the analyst should use a two-tailed t-test.

TYPES

The meaning of similarity depends on the type of data which has been collected. "Types" refers to the type of sampling conducted.

For instance, the degree of similarity of measurements one would expect from two random samples would be different from that one would expect from a before and after study on the same sample. With two random samples, a standard two-tailed t-test would be in order. However, in a before and after study on the same sample, the researcher might not only expect the average after measurement to be demonstrably higher (or lower) than the average before measurement, but also that each individual research subject's after measurement would be higher (or lower) than each individual research subject's before measurement.

When each measurement in one sample can be matched with a measurement in the second sample, the research holds paired data. In these situations, the paired type of t-test should be used.

The key determination in choosing to use a paired type of t-test is in the ability of the researcher to match each data point in one sample with a data point in the other sample. At times, the two samples do not have to contain the exact same research subjects, but rather research subject matching (though demographics or other means) may be used to pair the data. Furthermore, whether the researcher holds prior expectations with respect to the mean measurements has no impact on the choice of using a paired type t-test or not. In examining the differences between the samples with paired data, the research simply expects that any overall differences in distributions from the samples would also be reflected in the specific differences between the paired data. Thus, the paired type of t-test can be used both with one-tailed examinations and two-tailed examinations.

Besides the paired data, there are also two-sample homoscedastic data, and two-sample heteroscedastic data. **Homoscedastic data is data that is randomly distributed about the mean. Heteroscedastic data is data in which the variance about the mean systematically changes.** In the remainder of this text, we will assume the data is homoscedastic unless specifically directed otherwise.

T-TESTS IN EXCEL

Excel enables analysts to conduct a variety of t-tests. The Excel function TTEST(Sample A, Sample B, tails, type) returns the p-value of a t-test, or the probability that the differences between the samples arise from random sample error alone. Sample A and Sample B are arrays in a spreadsheet which contain the two sample sets of data to be compared. (The order of which sample is referred to as A versus B doesn't matter in conducting a t-test). Tails is either 1 or 2, depending on the number of tails needed to be evaluated. Type is either 1, 2, or 3, depending on the type of data being examined: (1) paired; (2) two-sample, homoscedastic; or (3) two-sample, heteroscedastic. In the majority of studies, the research will choose a two-tailed homoscedastic t-test, and will thus input TTEST(Sample A, Sample B, 2,2).

Excel Encoding	Data Type
1	Paired
2	Two-sample, homoscedastic
3	Two-sample, heteroscedastic

For example, a researcher wants to determine the likelihood of purchasing premium softened toilet paper and asks samples the following survey question measured on a 1 to 5 Likert scale:

How likely are you to purchase softened toilet paper if it is priced at a slight premium to standard toilet paper?					
Response Items	Highly Unlikely	Somewhat Unlikely	Neither Likely nor Unlikely	Somewhat Likely	Highly Likely
Encoding	1	2	3	4	5

From four different samples, the researcher collects the following data.

	A	B	C	D
1	Sample A	Sample B	Sample C	Sample D
2	2	4	4	3
3	2	3	3	2
4	4	4	1	4
5	1	2	2	2
6	4	4	2	2
7	4	4	2	4
8	4	4	1	1
9	3	2	3	2
10	5	5	3	3
11	3	5	1	4
12	3	5	2	3
13	5	1	2	1
14	3	3	2	3
15	1	2	2	4
16	4	4	1	5
17	2	1	3	1
18	2	3	1	4
19	1	3	2	4
20	3	2	1	3
21	3	3	2	1
22	2	3	5	3
23	5	5	4	2
24	3	3	2	5
25	4	1	1	2
26	2	2	1	5

The researchers would like to compare the measurements from Sample A with those in Sample B, and those from Sample C to those in Sample D. To compare these samples, the researcher directs the analyst to conduct a graphical analysis and statistical analysis comparing the survey results. The analyst will interpret these directives as

- Use the Data Analysis Histogram dialogue to conduct a frequency analysis of the survey results and find, for each sample:
  - the probability of obtaining a specific response.
  - the sample mean.
- Use the frequency analysis to plot a column chart of the distributions to be compared on the same plot.
- Conduct a t-test with two-tails and assume two-sample homoscedastic data.

  - Use hypothesis testing to compare the p-value of a t-test to a chosen significance level,
  - To use the 5% significance level, that is, set $\alpha = 0.05$,
  - To make statements regarding the differences or similarity of the datasets,

Sample A & B

In comparing Sample A and B, the analyst identifies the following results:

- Sample A has a mean response of 3.0, or "Neither Likely nor Unlikely".
- Sample B has a mean response of 3.1, or "Neither Likely nor Unlikely".
- The frequency of responses is shown below.
- Use the formula "=TTEST(A2:A26,B2:B26,2,2)"
  - to select the data arrays of responses from Sample A (B2:B26) and from Sample B (C2:C26),
  - to use a two-tailed t-test (2), and
  - to assume the data is homoscedastic (2).
- The p-value of student's t-test is 0.75, hence...
  - The likelihood that differences between Sample A and B could arise from random sample error is 75%.
  - The p-Value of 0.75 is greater than the chosen of α of 0.05.
  - We cannot reject the null hypothesis at the 5% significance level, thus we accept the null hypothesis.
  - The difference between the two sample measurements is due to random sample errors. AND...
  - The differences are not statistically significant.
- Thus, both Sample A and Sample B are equally likely to purchase softened toilet paper, and both samples report that they are Neither Likely nor Unlikely to purchase, that is, they scored a 3 on a 1 to 5 scale of likelihood.

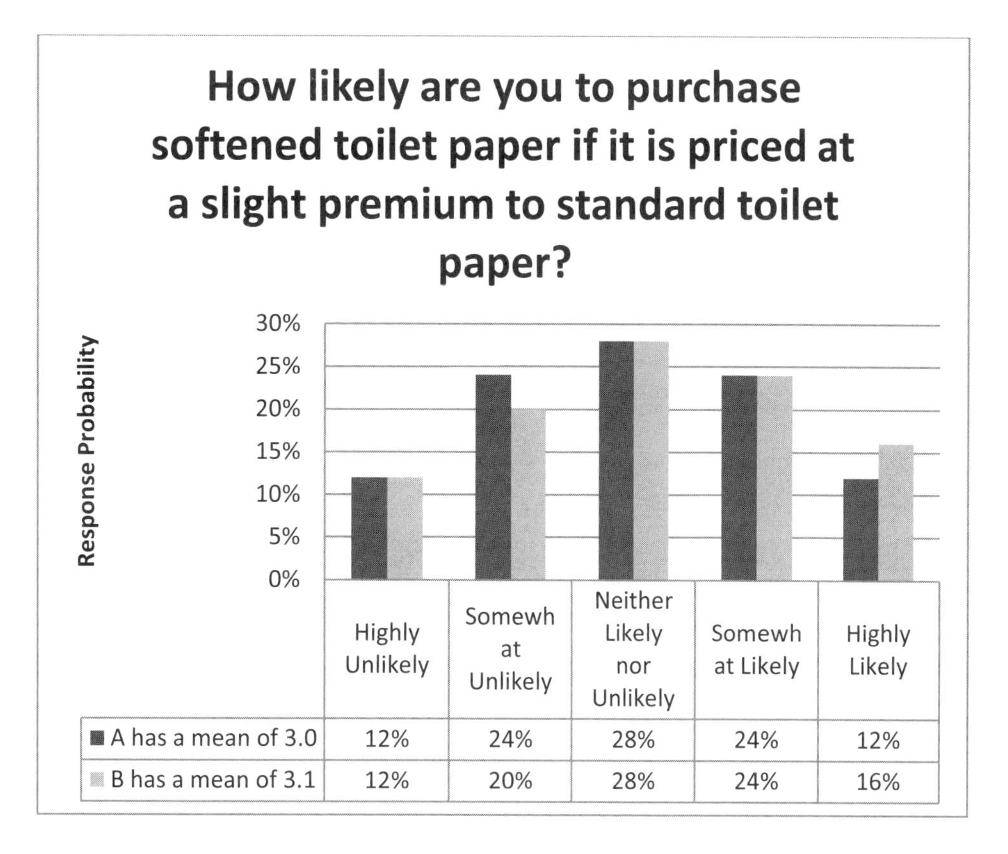

If Sample A consists of individuals who identify themselves as male and Sample B consists of individuals who identify themselves as female, the researcher could conclude that gender has little to do with the likelihood of purchasing softened toilet paper.

Sample C & D

We can repeat the process to compare Sample C and D.

- Sample C has a mean response of 2.1, or “Somewhat Unlikely”.
- Sample D has a mean response of 2.9, or “Neither Likely nor Unlikely”.
- The frequency of responses is shown below.
- Use the formula “=TTEST(C2:C26,D2:D26,2,2)”
    - to select the data arrays of responses from Sample C (C2:C26) and from Sample D (D2:D26),
    - to use a two-tailed t-test (2), and
    - to assume the data is homoscedastic (2).
- The p-value of student’s t-test is 0.02, hence...
    - The likelihood that differences between Sample C and D could arise from random sample error is 2%.
    - The p-value of 0.02 is less than the chosen of α of 0.05.
    - We can reject the null hypothesis at the 5% significance level, and thus we accept the alternative hypothesis.
    - The difference between the two sample measurements arises from the samples representing two different populations. AND ...
    - The differences are statistically significant.

- Thus, Sample C is less likely to purchase softened toilet paper than Sample D.

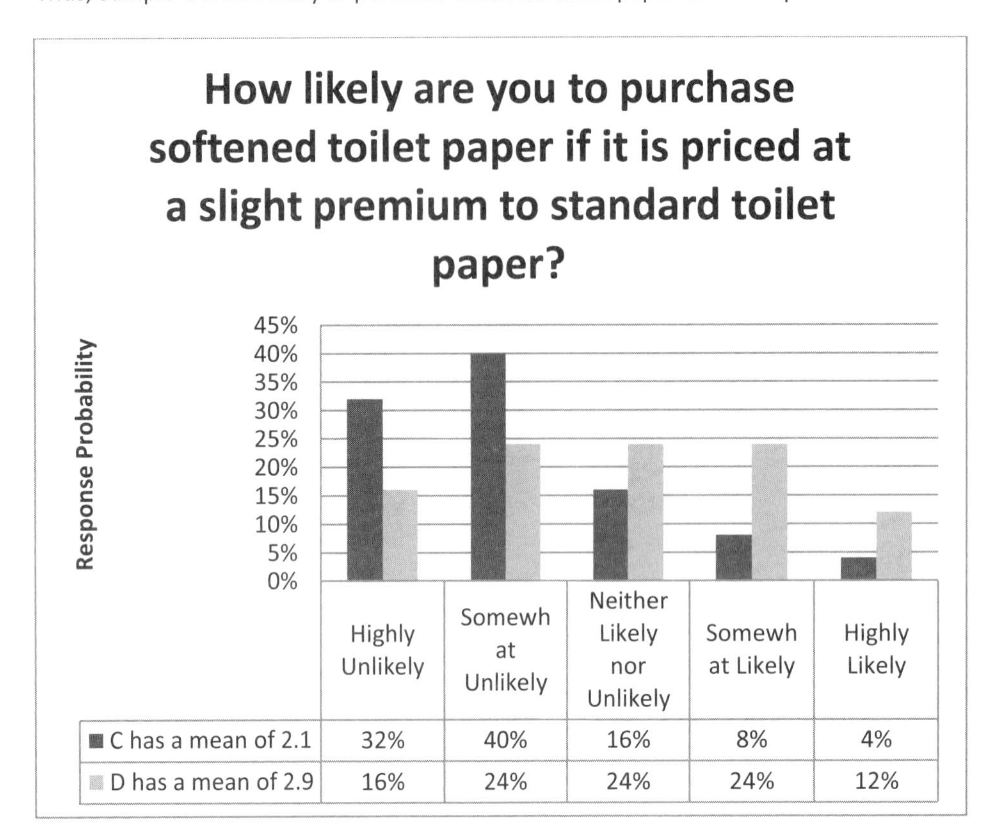

	Highly Unlikely	Somewhat Unlikely	Neither Likely nor Unlikely	Somewhat Likely	Highly Likely
C has a mean of 2.1	32%	40%	16%	8%	4%
D has a mean of 2.9	16%	24%	24%	24%	12%

If Sample C consists of individuals who support the Green Party and Sample D consists of individuals who support the Democratic or Republican Party, the researcher could conclude that political party affiliation influences the likelihood of purchasing softened toilet paper.

Marketing Metrics: Demand Estimation

How large is a market? If there are 300 million Americans, is the relevant market size in America always 300 million? That is, should every marketer expect every single American will purchase their product? Men and women alike? Rich and poor alike? Babies and retirees alike? That would be a ridiculous claim, especially for sales of something like heavy earth moving machinery.

Marketers need to be able to estimate the demand of a market. Demand estimates inform production strategy, distribution requirements, shelf space, promotional intensity, sales force size, and almost every other dimension of a company. To estimate demand, marketers will collect the relevant facts available, make a few assumptions concerning how the facts are related, and use the assumptions and facts to calculate demand estimates.

Many times, the assumptions can be expressed in the form of ratios. Demand estimates generated with ratio expressions of assumptions are often called the "chain ratio method". The chain ratio method is highly flexible and can be used in a number of situations.

Consumer Markets

Suppose a marketer seeks to estimate demand for his brand of grape soda. He knows the following two facts: (1) the market consists of 291 million people; (2) a person in that market buys 3 sodas a month on average.

With these facts, the marketer can now estimate the overall annual demand for sodas using the chain ratio. Expressing the average demand as a ratio and the relationship between months and years as a ratio, we find from the chain ratio method the following facts.

$$\frac{3\ sodas}{person - month} \cdot \frac{12\ months}{year} \cdot 291\ million\ people = \frac{10.5\ billion\ sodas}{year}$$

The marketer has now determined that the overall market for sodas is 10.5 billion sodas per year. (To convert millions into billions, recall one million equals 1,000,000 and one billion equals 1,000,000,000.)

Yet, the marketer doesn't expect to take the entire market. In fact, the overall market for sodas may be broken down into categories and market shares of colas (55%), lemon-lime (30%), and specialty (15%). Furthermore, grape sodas only constitute 10% of the sales of specialty sodas overall, and the marketer is only expecting a 30% share of the overall grape soda market.

Given these expectations, the marketer can continue using the chain ratio method to estimate demand for his brand of sodas.

$$\frac{10.5\ billion\ sodas}{year} \cdot \frac{15\%\ specialty\ sodas}{all\ sodas} \cdot \frac{10\%\ grape\ soda}{all\ specialty\ sodas} \cdot \frac{30\%\ brand\ market\ share}{all\ grape\ soda\ market} = \frac{47.3\ million\ branded\ grape\ sodas}{year}$$

The marketer can therefore estimate sales of his branded grape soda at 47.3 million units per year.

In general, the more specific a marketer can be in identifying their target market, the more accurate their forecast of demand will be. Moreover, by clearly defining the target market, the marketer can better define the product strategy, pricing strategy, placement strategy, and promotion strategy necessary to profitably capture their target market.

Sales Force Requirements

In pharmaceutical sales, it is often necessary to estimate sales force size requirements. A pharmaceutical formulary firm may know the number of patients with a given disease, average number of patients per doctor, and average number of salespeople required to interact with the doctors. With these facts, the marketer can estimate the required sales force.

Given the following metrics

Estimated Patients with Given Disease	37.1 million
Patients / Specialized Doctor	1,280
Doctors / Salesperson in a quarter	720

the estimated sales force requirement can be calculated to be

$$37.1\ million\ patients \cdot \frac{1\ doctor}{1280\ patients} \cdot \frac{1\ salesperson}{720\ doctors} = 40.3\ salepeople$$

From the facts, the pharmaceutical firm may determine that they need to prepare 40 salespeople to launch the new formulary.

Location Identification

Ratios can be useful for identifying locations for stores using US Census data.

For instance, from the US Census Bureau, we learn there are 309.9 million people in the US. From the US Economic Census, we can also learn there are 26,093 dry cleaners in the US, earning a total revenue of $8.09 billion. (Search for NAICS 812320, Dry cleaning and Laundry Services, on the American FactFinder for more current data).

Suppose a new high-rise condominium is being built and they are seeking an individual to open a dry cleaner. The condominium building will have 750 units with an average of 2.5 people per unit. Would locating dry cleaners in the building be a good marketing strategy?

To evaluate the soundness of this strategy, we can use the chain ratio method to (1) predict the revenue of a dry cleaner in the condominium building and (2) index the demand in that building against the national average.

Predicted Revenue

$$\frac{\$8.09\ billion\ Dry\ Cleaner\ Revenue}{309.9\ million\ People} \cdot 750\ units \cdot \frac{2.5\ people}{unit} = \$43{,}900\ Dry\ Cleaner\ Revenue$$

Indexed Demand

$$\frac{750\ units}{1\ Dry\ Cleaner} \cdot \frac{2.5\ people}{unit} \cdot \frac{26{,}093\ Dry\ Cleaners}{309.9\ million\ people} \cdot 100 = 16$$

From these ratios we discover that the Dry Cleaners can be expected to earn $43,947 in revenue, and that demand for a building-specific dry cleaners should be 16% of that found for a dry cleaner's establishment on average. Hence, opening dry cleaners to serve a single building is a risky prospect. Unless the people in that condominium building use dry cleaners six times more than the average American, the dry cleaner will suffer a below average revenue. If the building is full of white collar workers, this might be a realistic expectation. If the building is full of blue collar workers, the marketer may want to pass on the opportunity.

Time Scales

Another use of the chain ratio method is in translating purchase frequency data evaluated on one time scale to that of another time scale.

For instance, in calculating category incidence frequencies in Chapter 1, we assumed that all responses were provided using the same unit of time, such as days, weeks, months, OR years. At times, data will be collected in mixed units of time, such as days, weeks, months, and years. When working with mixed time scales, we must convert each interpretation of the response items into a common metric of time.

A straightforward approach is to convert all time scales to the common unit of 1 year, identify the category incidence frequency on the annual basis using weighted averages, and convert this annual category incidence frequency to the period of time desired by the researcher.

For example, suppose an entrepreneur is considering opening a neighborhood roast chicken outlet and estimates that the immediate neighborhood has 12,500 people, that the average roasted chicken sells for $9.00, and that the vendor only plans on being open 6 days a week. He has surveyed people in the neighborhood regarding their anticipated frequency of demanding roasted chicken. He desires to estimate demand and revenue on a monthly basis, as his lease will be due on a monthly basis.

After conducting a survey, the entrepreneur may have gained the following responses:

Roasted Chicken Eating Frequency	Response Frequency	Response Probability
Once a Year	8	4%
Once a Month	58	29%
Twice a Month	92	46%
Once a Week	36	18%
Twice or More a Week	6	3%

This research includes weeks, months, and years. We can interpret each response item as follows:

- Once a year implies 1 sale / year.
- Once a month implies 12 sales / year, since there are 12 months in a year.
- Twice a month implies 24 sales / month, since there are 2 sales per month and 12 months in a year.
- Once a week implies 52 sales / year since there are 52 weeks in a year.
- Twice or more a week implies at least 2 sales / week but could be interpreted at 3 sales / week, 4 sales / week, or even 7 sales per week. **In conducting estimates of demand, researchers usually choose conservative interpretations of data rather than aggressive interpretations. As such, the researcher would reject the interpretation of "Twice or More a Week" to mean 7 sales per week. Instead, they will conservatively interpret the demand at the lower bound of 2 sales per week. Since there are 52 weeks in a year, 2 sales per week implies 104 sales per year.**

Hence, the response items can be interpreted as:

Roasted Chicken Eating Frequency	Annual Incidence Frequency	Response Probability
Once a Year	1	4%
Once a Month	12	29%
Twice a Month	24	46%
Once a Week	52	18%
Twice or More a Week	104	3%

Now, weighting the annual category incidence responses by the response probability, we calculate the weighted average category incidence at 27.0 chickens per year. Converting annual category incidence to monthly category incidence requires dividing by 12. The entrepreneur uses this conversion ratio to determine an expected demand per person of 2.25 chickens.

Finally, using the chain ratio to estimate neighborhood demand in both units and revenue, the entrepreneur calculates

$$\frac{2.25\ chickens}{person - month} \cdot 12{,}500\ people = \frac{28{,}100\ chickens}{month}$$

$$\frac{28{,}100\ chickens}{month} \cdot \frac{\$9.00}{chicken} = \frac{\$253{,}000}{month}$$

From this calculation, the entrepreneur anticipates sales of 28,100 chickens per month, or $253,000 in revenue per month.

Exercises

Grocery Targets

A large retail grocery store wanted to target its customers. They decided to see whether purchases of morning customers differed significantly from those of non-morning customers. The dollar amount of purchases in the morning and non-morning by each group are shown below. Samples were randomly selected from the store's database.

DOLLAR SPENT BY CUSTOMERS IN PAST YEAR	
Morning Customers	Non-Morning Customers
42	20
20	21
18	12
16	19
22	15
35	16
17	14
21	11
19	16
25	17
20	27

1. Calculate the average dollars spent for each group, morning and non-morning customers.
2. Using 5 intervals for all the data, identify the frequency distributions of dollars spent for both samples. (Hint: use the minimum of both samples and the maximum of both samples to identify the appropriate common set of intervals to apply to both samples).
3. On the same plot, chart the distributions.
4. Conduct a t-test to determine the significance of the differences between sample measurements.
    a. How many tails should be used in the t-test?
    b. What type of data is this: paired, homoscedastic, heteroscedastic?
    c. What is the p-value of the t-test?
    d. What is the likelihood that random sample error alone could explain the differences in the measurements?
5. At the 5% significance level
    a. Is the p-value greater than or less than α?
    b. Should the null hypothesis be accepted or rejected?
    c. Are the differences between the sample measurements statistically significant?
    d. Is the difference due to random sample error or are the samples taken from different populations?
6. In plain English, give a complete, one sentence interpretation of the result appropriate for managerial decision making.

NEW TRAINING PROGRAM

At the Smith-Williams Manufacturing Company, new employees are expected to attend a 3-day seminar to learn about the company. At the end of the 3 days, they are tested to measure their knowledge about the company. Traditionally, the training method consisted of lectures accompanied by question and answer sessions. Management experimented with a different training method, which consists of having new employees view videocassettes for 2 days without any question and answer session. If the new system works as well as the traditional one, it will save the company a lot of money over a period of several years. The company managers want to know if there is a significant difference in the effectiveness of the two training sessions.

Assume managers "randomly" sample one group of 15 new employees to train traditionally and another group of 15 new employees to be trained by the 2-day videocassette method. Assume the scores for the test are normally distributed and that the population variances are approximately equal. Choose the .05 alpha (significance) level for the t-test.

TEST SCORES FOR NEW HIRES									
Traditional Method					New Method				
52	44	52	47	47	59	54	55	65	52
48	42	51	42	43	57	64	53	53	56
56	50	53	45	48	53	57	58	62	57

1. Calculate the average test score for both samples.
2. Using 5 intervals for all the data, identify the frequency distributions of test scores and chart the distributions on the same plot.
3. Conduct a t-test to determine significance of the differences between sample measurements.
    a. What is the p-value of the t-test?
    b. What is the likelihood that random sample error alone could explain the differences in the measurements?
    c. Is the p-value greater than or less than α?
    d. Should the null hypothesis be accepted or rejected?
    e. Are the differences between the sample measurements statistically significant?
    f. Is the difference due to random sample error or are the samples taken from different populations?
4. In plain English, give a complete, one sentence interpretation of the result appropriate for managerial decision making.

ATKINS DIET

A researcher was interested in learning if following the Atkins diet resulted in greater weight loss than following the Weight Watchers diet. Twenty persons who had made New Year's resolutions to diet were randomly assigned to one of two treatment conditions: Atkins (N = 10) and Weight Watchers (N = 10). The dieters were instructed to follow the rules of the diet to which they were assigned for a six-week period. The number of pounds each dieter lost is represented below.

Weight Loss in Pounds									
Traditional Method					Weight Watchers				
10	16	17	17	14	12	11	13	11	15
14	12	13	15	9	10	15	18	7	8

The analyst is directed to choose the appropriate t-test and significance level.

1. Calculate the average weight loss for both samples.
2. Using 5 intervals for all the data, identify the frequency distributions of weight losses and chart the distributions on the same plot.
3. Conduct a t-test to determine significance of the differences between sample measurements.
    a. What is the p-value of the t-test?
    b. What is the likelihood that random sample error alone could explain the differences in the measurements?
    c. Is the p-value greater than or less than α of 0.05?
    d. Should the null hypothesis be accepted or rejected?
    e. Are the differences between the sample measurements statistically significant?
    f. Is the difference due to random sample error or are the samples taken from different populations?
4. In plain English, give a complete, one sentence interpretation of the result appropriate for managerial decision making.

Graduates vs. Non-graduates Purchasing

A large bookstore chain wanted to target its customers. They decided to see whether purchases of college graduates differed significantly from those of non-college graduates. The number of books purchased annually by each group is shown below. Assume homogeneity of variances. Samples were randomly selected from the store's database.

NUMBER OF BOOKS PURCHASED IN LAST YEAR									
College Grads					Non-College Grads				
12	18	22	17	19	10	21	12	9	15
20	16	15	21	25	16	14	11	16	7
20	16				17	8			

The analyst is directed to choose the appropriate t-test and significance level.

Compare the mean books purchased for college grads to non-college grads using the 5% significance level and interpret the results. Provide a graphical analysis to support your managerial interpretation.

SALARIES

An experiment is conducted to compare the starting salaries of male and female college graduates who find jobs. Pairs are formed by choosing a male and a female with the same major and similar grade point averages (GPAs). Suppose a random sample of 10 pairs is formed in this manner and the starting annual salary of each person is recorded. The results are shown in the table below.

Pair	Male	Female
1	$29,300	$28,800
2	41,500	41,600
3	40,400	39,800
4	38,500	38,500
5	43,500	42,600
6	37,800	38,000
7	69,500	69,200
8	41,200	40,100
9	38,400	38,200
10	59,200	58,500

1. Calculate average salary for both males and females.
2. Using 5 intervals for all the data, identify the frequency distributions of salaries for both samples. (Hint: use the minimum of both samples and the maximum of both samples to identify the appropriate common set of intervals to apply to both samples).
3. On the same plot, chart the distributions.
4. Conduct a t-test to determine significance of the differences between sample measurements.
   a. How many tails should be used in the t-test?
   b. What type of data is this: paired, homoscedastic, heteroscedastic?
   c. What is the p-value of the t-test?
   d. What is the likelihood that random sample error alone could explain the differences in the measurements?
5. At the 5% significance level:
   a. Is the p-value greater than or less than α?
   b. Should the null hypothesis be accepted or rejected?
   c. Are the differences between the sample measurements statistically significant?
   d. Is the difference due to random sample error or are the samples taken from different populations?
6. In plain English, give a complete, one sentence interpretation of the result appropriate for managerial decision making.

OIL

A major oil company wanted to estimate the difference in average mileage for cars using regular engine oil compared with cars using a synthetic-oil product. The company used a paired-sample approach to control for any variation in mileage arising from different cars and drivers. A random sample of 10 motorists (and their cars) was selected. Each car was filled with gasoline, the oil was drained, and new, regular oil was added. The car was driven 200 miles on a specified route. The car was then filled with gasoline and the miles per gallon were computed. After the cars completed this process, the same steps were performed using synthetic oil.

Pair	Synthetic oil	Regular oil
1	19.8	20.7
2	28.8	25.8
3	20.4	27.8
4	18.7	14.9
5	23.4	21.6
6	27.1	21.1
7	28.4	28.0
8	21.4	13.0
9	26.4	24.4
10	19.9	14.3

Compare the mean mileage for both synthetic and regular oil users using the 5% significance level and interpret the results. Provide a graphical analysis to support your managerial interpretation.

Job Satisfaction

A company institutes an exercise break for its workers to see if this will improve job satisfaction, as measured by a questionnaire that assesses workers' satisfaction. Scores for 10 randomly selected workers before and after implementation of the exercise program are shown. The company wants to assess the effectiveness of the exercise program.

	Job Satisfaction Score	
Worker	Before	After
1	34	33
2	28	36
3	29	50
4	45	41
5	26	37
6	27	41
7	24	39
8	15	21
9	15	20
10	27	37

Compare the mean job satisfaction for both before and after the implementation of exercise breaks using the 5% significance level and interpret the results. Provide a graphical analysis to support your managerial interpretation.

Optical Scanning Market Estimation

Calculate the market potential for sales of special optical scanning systems. For each million in sales that a company does, 35 systems can be sold. There are 25 prospective companies with 5 million each in sales, and 15 prospective companies with 10 million each in sales.

Estimate sales performance in a territory based on industry numbers.

Coin Operated Laundry Facility

An entrepreneur is interested in setting up a coin operated laundry facility in a town of 30,000. The NAICS is 812310, Coin-Operated Laundries.

1. On the US Economic Census website, look up
    a. The total revenue earned by Coin-Operated Laundries in the latest reporting year.
    b. The number of Coin-Operated Laundries in the latest reporting year.
    c. What is the latest reporting year?
2. Calculate the average revenue earned per coin operated Laundromat.
3. Calculate the average revenue earned per citizen if we assume 309.9 million Americans.
4. How much revenue should the entrepreneur expect to earn if he establishes the only coin operated Laundromat in town?
5. What would his indexed demand be?
6. Name at least two other facts you would like to collect before supporting this entrepreneurial endeavor.

Starbucks finds a Target Market

A researcher hypothesized that older persons were not as good customers for Starbucks as were younger persons. The data are shown below.

	Respondents	
Purchase Frequency	Under 25	25 years old or older
Once a week or more	40	7
Once a month to once a week	51	32
Once a year or less	8	54

1. Interpret the response categories using a conservative interpretation for demand.
2. What is the weighted average number of coffees an under 25 person purchases per year? Per month? Per week?
3. What is the weighted average number of coffees a 25 or older person purchases per year? Per month? Per week?
4. What is the mean of the weighted average number of coffees sold to both the under and over 25 year olds?
5. If the average coffee sold at Starbucks is priced at $2.75 and has an 80% contribution margin, what is the annual profit earned per
    a. Under 25 years old customer?
    b. 25 years old or older customer?
6. Which market segment looks to be more profitable?

CHAPTER 7: ANOVA /F-TESTS

While t-tests are useful for comparing numerical data between just two samples, there is another test that is more appropriate when reviewing multiple numerical samples.

F-tests, also called an analysis of variance, or ANOVA, enable analysts to investigate the differences and similarity of more than two sets of data with a single statistic. F-test is used with numerical data collected from different categorical groups. For example, scores on a test, mileage under different driving conditions, or number of cups of coffee purchased per person.

Researchers often deal with several samples at a time, and sometimes samples are examined in multiple dimensions. For instance, data might be collected from numerous people and that data might be divided into two samples according to gender (male or female) in one factor or dimension and at the same time divided into three samples according to age groups (under 19, between 19 and 65, and over 65) in another factor or dimension. As the number or samples increases, the number of t-tests required to analyze the data increases exponentially. Therefore, the potential for introducing statistical error in the analysis increases exponentially. In these cases, an analyst will instead conduct an F-test rather than several t-tests.

While the F-Test and t-test each uses a different calculation, the analyst can rest easy knowing that each test results in a similar metric useful for interpreting the results: the p-value. The p-value from an F-test can be compared to the alpha in a test of significance in the exact same manner as that used in a t-test.

To discuss the theoretical background of an F-test, we will introduce the concepts of dependent versus independent variables, degrees of freedom associated with calculating a statistic, and the approach to conducting an analysis of variance.

In this chapter, students will learn:

- How to conduct and interpret an F-test with a One-Way ANOVA using nominal independent variables and numerical dependent variables.
- How to conduct and interpret an F-test with a Two-Way ANOVA using nominal independent variables and numerical dependent variables.
- How to calculate the Return on Investment for some types of products.

VARIABLES

When investigating the relationship between groups and a research metric, the analyst will have independent variables and dependent variables.

For example, suppose the enjoyment of coffee was measured using a one to five Likert scale according to level of agreement with the statement: "Coffee is an enjoyable beverage." Furthermore, suppose research respondents were identified by gender and race. Would different genders enjoy coffee at different levels? Would different races enjoy coffee at different levels?

Each of the above questions share common format: the researcher is examining the effect of an independent nominal variable on a numerical variable. The independent nominal variables are gender and race. The dependent variable is the level of agreement. The researcher is searching for changes in the independent variable that influence a change in the dependent variable. **Independent variables are variables which inherently vary among the sample. Dependent variables are variables in which their variations are, in some way, driven by the variation of an independent variable.**

F-tests examine whether variations in an independent nominal variable influence changes in a dependent numerical variable.

For examples of variables used in an F-test, consider the following:

Nominal Independent Variables	Dependent
Gender	Coffee Interest
Ethnicity	Seafood Demand
Industry Classification	Willingness to Pay
Geographic Region	Demand

DEGREES OF FREEDOM

Degrees of freedom, often denoted by df, is the number of items in a calculation of a statistic that are free to vary.

When calculating the mean of a sample, the degrees of freedom is equal to the number of elements in the data set. For instance, the set of data [2, 4, 7] has three elements. Three different data-points could vary, and hence for calculating the mean of this sample data, the degrees of freedom is three. And, when calculating the mean of [2, 4, 7], we divide by 3 (see Chapter 2).

However, when calculating the sample standard deviation, the degrees of freedom is one less than the number of elements in the data set. Why "one less"? We will have used one degree of freedom in calculating the sample mean, and the sample mean is used in calculating the standard deviation, hence the degrees of freedom is reduced by one. For instance, the set of data [2, 4, 7] has three elements and the calculation of the mean uses one of these degrees of freedom, hence the calculation of the sample's standard deviation has only two degrees of freedom. And, when calculating the sample deviation of [2, 4, 7], we divide by 2=3-1 (see Chapter 2).

Familiarity with degrees of freedom will increase through the use of this concept in this and the remaining chapters.

THE F-TEST

The F-test examines the ratio of the variability between the groups to the variability within the groups.

$$F\ Statistic = \frac{Variance\ between\ Groups}{Variance\ within\ Groups}$$

- **If the variability between the groups is similar to the variability within the groups, then the ratio will be relatively small, and we will suspect that the different groups are relatively similar.**
- **However, if the variability between the groups is much larger than that which would have been predicted on the basis of the variability within the groups, then the researcher will suspect that the groups are relatively different. The F statistic, in this case, will be relatively large.**

How small is small and how large is large? As with the t-test, the F-test relies upon a distribution and a significance level: if the F statistic is greater than a critical value of the F distribution ($F_{critical}$) determined by the chosen significance level, then we would reject the null hypothesis and conclude that the different groups indeed have different measurements.

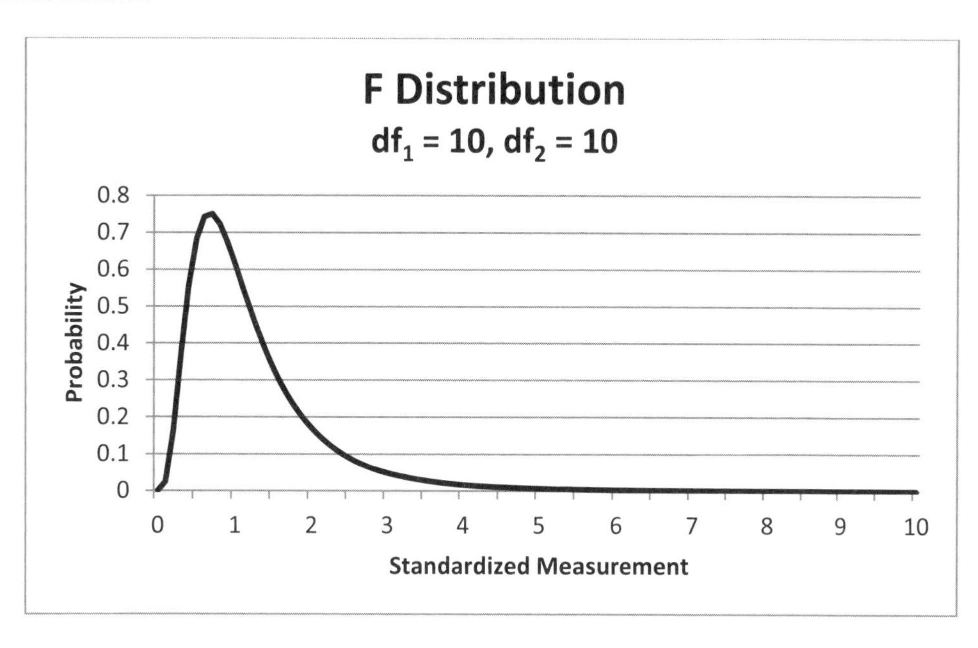

Fortunately, with modern analytical tools, we rarely need to calculate the $F_{critical}$. Modern analytical software will generate a p-value of an F-test for us. As with the p-value of a t-test, the p-value of an F-test can be compared with the chosen significance level (α).

- **If the p-value is greater than α:**
    - **we can accept the null hypothesis, and**
    - **conclude that the independent variable has no statistically significant effect on the dependent variable.**
- **If the p-value is equal to or less than α:**
    - **we would reject the null hypothesis, and**
    - **conclude that the independent variable has a statistically significant effect on the dependent variable.**

The F-test is very similar to a t-test, but they differ in use. **Although an F-test can be used with two samples, it is usually reserved for more than two samples. With only two samples, it is more parsimonious to use a t-test even though results from an F-test would be the same.** In contrast, if there are three or more samples and multiple t-tests would be needed to to uncover differences, there is always a greater probability of error in interpreting the resultant t statistic. As such, the F-test is appropriate for three or more samples.

Other points of comparison between t-tests and F-tests can be found in the chart below.

	t-test	F-test
Two samples at a time	Yes	No
More than two samples simultaneously	No	Yes
Assumes random sampling	Yes	Yes
Relies on numerical measurements of the observed dependent data	Yes	Yes
Relies on nominal categorization of the sample groups	Yes	Yes
Assumes the observations are normally distributed in the populations	Yes	Yes
Assumes both populations have similar but unknown sample standard deviations	Yes	Yes

To calculate the F statistic, we must discuss ANOVA.

ANOVA

ANOVA, or analysis of variance, examines the difference between group means. It does this by comparing the variance between the groups to the variance within the groups – as described above with the F statistic. ANOVA is used with numerical data, including both interval and ratio data.

The concept behind ANOVA can be described by the phrase "birds of a feather flock together." That is, if the birds are similar, differences between groups of birds are small and relatively not interesting to the analyst. Birds on the right hand side of the flock will be relatively interchangeable with birds on the left hand side of the flock. In contrast, if the groups of birds are dissimilar, the analysts will obtain a significant F-test statistic and the birds won't even flock together. **With respect to ANOVA, we assume that similar groups will have variances which are relatively meaningless but dissimilar groups will have variances much greater than the variance within the individual groups.**

For instance, suppose an analyst needs to compare and analyze differences in sales between three different groups of perfume shops. Out of a total of 15 perfume shops, 5 received a low level of promotional spending, 5 received a moderate amount of promotional spending, and 5 received a high amount of promotional spending. The analyst needs to compare how the independent variable in each group impacts the dependent variable in each group. In this example, the independent variable has three levels – low, moderate, and high promotional spending. The dependent variable is a numerical score of the mean sales for each group. Since there are multiple groups to compare to one other, the analyst will conduct an F-Test or an ANOVA. ANOVA examines the difference between two or more group means.

Types of ANOVAs

There are two basic types of ANOVAs which differ in terms of their samples, Between Groups ANOVA and Repeated Measures ANOVA.

In the Between Groups ANOVA, the samples are separate and independent of one another. Different treatment levels would be administered to different groups. The example of the perfume stores represents a between groups ANOVA because the analyst is measuring and comparing the effect of different levels on separate groups.

In the Repeated Measure ANOVA, the same sampling units are administered different treatment levels, but at different times. Taking the example from above, all 15 stores would have a low level of promotional spending for a set period of time, and then the same 15 stores would receive a moderate amount of promotional spending for the same amount of time, and finally the same 15 stores would receive a high amount of spending. The mean sales would be recorded for each level of promotional spending and compared to one another.

In summary, the between groups ANOVA tests separate independent sampling units, whereas the repeated measures ANOVA test the same sampling units over and over again.

One-Way and Two-Way ANOVAS

As we did with a t-test, a One-Way ANOVA looks at group differences using one variable or one "factor." The perfume store example would be an example of a One-Way ANOVA because there was only one variable – the amount of promotional spending. ANOVA is used rather than a t-test due to the need to compare more than two samples.

One of the beauties of ANOVA is that more than one variable, or more than one "factor," can be analyzed at a time. Using the perfume store example, one could examine the effects on mean sales of two variables at the same time: promotional spending and location of stores (either in an outlet mall or a regular mall). Thus, a significant difference between means under different levels of promotional spending could be obtained, referred to as a "main effect" for promotional spending. A significant difference could also be obtained between locations referred to as a "main effect" for locations. It could be that no significant main effect was obtained for either of the two variables, but rather that the variables interact to yield a significant "interaction effect." Perhaps regular mall locations perform better under high level of spending than do outlet malls. In contrast, the outlet malls may perform better under low levels of spending than do regular malls.

ANOVA Calculation

As mentioned, ANOVA compares variances between groups to variances within groups. To calculate the F statistic, we must calculate a number of intermediate statistics.

First we calculate the mean and variance of each group as well as the grand mean. (Variance is simply the standard deviation squared). Continuing with our simple promotional spending example for perfume stores and sales, we would have three different mean sales scores and three different variances; one for each level of promotional spending. Using the index K to represent the different groups, where the Kth group contains the X_{Ki} data-points and has N_K elements, we find the mean and variance of the Kth group to be

$$Mean\ of\ the\ K^{th}\ Group = \bar{X}_K = \frac{\sum_{i=1}^{N_k} X_{Ki}}{N_K}$$

$$Variance\ of\ the\ K^{th}\ Group = s_k^2 = \frac{\sum_{i=1}^{N_k} (X_{Ki} - \bar{X}_K)^2}{N_K - 1}$$

We can similarly calculate the grand mean, or the average sales across all three samples

$$Grand\ Mean = \bar{X}_G = \frac{\sum_{k=1}^{K} \bar{X}_k}{K}$$

where K is the number of groups, or 3 in the perfume example.

Second, we can calculate what is colloquially called the sum of squares, both within groups and between groups. The sum of squares is a measure of the variation. Within groups, we would calculate

$$Sum\ of\ Squares\ within\ Groups = SS_w = \sum_{k=1}^{K}(N_k - 1)s_k^2$$

And across groups, we would calculate

$$Sum\ of\ Squares\ between\ Groups = SS_b = \sum_{k=1}^{K} N_K(\bar{X}_k - \bar{X}_G)^2$$

Third, we would calculate the degrees of freedom associated with each of the measures of the sum of squares. In our perfume example, there are fifteen stores grouped into three groups. Hence, the total number of stores is N=15 and the number of groups is K=3. In calculating the degrees of freedom within groups, we use df_b=15-3=12, because there are fifteen data points and at least three degrees of freedom have been used in calculating the individual sample means.

$$Degrees\ of\ Freedom\ within = df_w = N - K$$

In calculating the degrees of freedom between the groups, we use df_b=3-1=2, as there are three groups and one of the degrees of freedom between the groups has been used in calculating the grand mean.

$$Degrees\ of\ Freedom\ between = df_b = K - 1$$

Fourth, we calculate the means squares, both within and between sample groups. The mean squares statistic is the sum of the squares statistic divided by the relevant degrees of freedom. As such, the mean squares statistic is a normalized measure of variance. The following two equations describe the Mean Squares within and between groups.

$$Mean\ Squares\ within = MS_w = \frac{SS_w}{df_w}$$

$$Mean\ Squares\ between = MS_b = \frac{SS_b}{df_b}$$

Fifth and finally, we calculate the F statistic. As mentioned above, the F statistic is a ratio of the variance between groups to the variance within groups. Because the mean squares statistic is a normalized measure of variance, we can calculate the F statistic simply as the ratio of the mean squared between groups to the mean squares within groups.

$$F\ Statistic = \frac{Variance\ between\ Groups}{Variance\ within\ Groups} = \frac{MS_b}{MS_w}$$

The analyst can either compare the F statistic to the critical F at a given significance level, or allow the modern statistical software to calculate the p-value for direct comparison to the chosen significance level.

F-Test and Excel

With Excel, an analyst can easily perform an F-test.

One-Way ANOVA

Returning to our example of a perfume retailer experimenting with different levels of promotion, suppose that the following data was collected from a sample of 15 different stores with three different levels of promotion:

	A	B	C
1	Low	Medium	High
2	$ 18,808	$ 18,700	$ 24,559
3	$ 17,308	$ 23,144	$ 22,362
4	$ 16,308	$ 18,461	$ 24,050
5	$ 19,308	$ 20,950	$ 19,424
6	$ 17,808	$ 20,320	$ 23,853

After placing this data into a blank spreadsheet, we run a one-way ANOVA.

- On the "Data" tab, clicking on "Data Analysis", select "Anova: Single Factor" for a one-way ANOVA.
- In the Anova Single Factor dialogue box:
    - Select the input range to be the entire range of data to be analyzed, cells A2:C6.
    - Leave the alpha at 0.05 for the 5% significance level.
    - Choose to put the Output Range in a blank cell.
    - Hit "ok".
- The following table should be generated.

Anova: Single Factor

SUMMARY

Groups	*Count*	*Sum*	*Average*	*Variance*
Column 1	5	89538.46	17907.69	1425000
Column 2	5	101574.2	20314.84	3613210
Column 3	5	114248.2	22849.65	4335792

ANOVA

Source of Variation	*SS*	*df*	*MS*	*F*	*P-value*	*F crit*
Between Groups	61070869	2	30535435	9.77238	0.003031	3.885294
Within Groups	37496007	12	3124667			
Total	98566876	14				

In the first half of the output are a number of summary statistics. Specifically, we find the means for each group in the Average column, and the variance for each group in the Variance column. The different groups are denoted as Column 1 for Low, the first column of data; Column 2 for Medium, the second column of data; Column 3 for High, the third column of data.

In the second half of the output, SS is the sum of squares, df is the degrees of freedom, MS is the mean squares, F is the F statistic, and F_{crit} is the critical F statistic.

Most importantly, the second half of the output contains the p-value. The p-value for these samples is 0.003, or 0.3%. Since the p-value is less than the 5% significance level, we can make the following deductions:

- Reject the null hypothesis.
- Accept the alternative hypothesis.
- The difference in sample means is statistically significant.
- The different samples arise from different populations.
- The level of promotion influences the sales.
    - A low level of promotion yields $17,900 in sales per week on average.
    - A medium level of promotion yields $20,300 in sales per week on average.
    - A high level of promotion yields $22,800 in sales per week on average.
- Higher levels of promotions tend to yield higher levels of sales.

Two-Way ANOVA

A two-way ANOVA assumes that the data has been gathered from samples which varied in two different dimensions rather than simply one. For instance, building on our example of a perfume retailer experimenting with different levels of promotion, suppose that the retailer also varied the color of the promotion between Aqua and Purple and collected the following data from 18 different stores.

	A	B	C	D
1		Low	Medium	High
2	Purple	$ 18,808	$ 18,700	$ 24,559
3		$ 17,308	$ 23,144	$ 22,362
4		$ 16,308	$ 18,461	$ 24,050
5	Aqua	$ 19,308	$ 20,950	$ 19,424
6		$ 17,808	$ 20,320	$ 23,853
7		$ 16,108	$ 18,996	$ 21,046

After placing this data into a blank spreadsheet, we run a two-way ANOVA.

- On the "Data" tab, clicking on "Data Analysis", select "Anova: Two Factor with replication" for a two-way ANOVA.
- In the "Anova Two-Factor" dialogue box:
    - Select the input range to be the entire range of data to be analyzed including the column headers and row headers, cells A1:D7.

  - Input "3" for Rows per sample, because there are three of each of the different colors of promotion.
  - Leave the alpha at 0.05 for the 5% significance level.
  - Choose to put the Output Range in a blank cell.
  - Hit "ok".
- The following table should be generated.

The first three parts of the output describe the averages under different conditions (Low, Medium and High promotion, Purple or Aqua promotion, and totals).

The fourth part of the output provides the relevant F statistic and p-values.

- The p-value for Sample is 0.44, or well above our significance level of 5%. Hence, the main effect of the type of store, either Aqua or Purple, which Excel has named the "Sample" variable since it is identified by the row grouping, has no statistically significant effect on the sales level.
- The p-value for Columns is 0.001, well below our significance level of 5%. Hence, the main effect of the level of promotion which Excel has named the "Columns" variable, since it is identified by the column, has a statistically significant effect on the sales level.
- The p-value for Interaction is 0.43, well above our significance level of 5%. Hence, the interaction effect between the level of promotion and the color of promotion has no statistically significant effect on the sales level.

Anova: Two-Factor With Replication

SUMMARY	Low	Medium	High	Total
Purple				
Count	3	3	3	9
Sum	52424	60305	70971	183700
Average	17475	20102	23657	20411
Variance	1583333	6956124	1322539	9685845
Aqua				
Count	3	3	3	9
Sum	53224	60266	64323	177813
Average	17741	20089	21441	19757
Variance	2563333	994665	5021029	4773046
Total				
Count	6	6	6	
Sum	105648	120571	135294	
Average	17608	20095	22549	
Variance	1680000	3180367	4010624	

ANOVA

Source of Variation	SS	df	MS	F	P-value	F crit
Sample	1925376	1	1925376	0.626443	0.444027	4.747225
Columns	73241554	2	36620777	11.91499	0.001411	3.885294
Interaction	5547528	2	2773764	0.902476	0.431398	3.885294
Within	36882049	12	3073504			
Total	1.18E+08	17				

Marketing Metrics: Return on Investment

In business-to-business sales, customers are interested in knowing the impact of a purchase on their firm. While a product may impact a customer in many ways, one of the key impacts customers will examine is the financial impact. The financial impact includes both costs required to acquire the product and the improvements to profitability that the product will deliver. To include both costs and improvements to profitability in a single metric, customers will want to know the return on investment.

The return on investment, or ROI, is a ratio of the financial gains after paying for the item to the cost of the item.

$$\boldsymbol{ROI = \frac{Financial\ Gain - Cost}{Cost}}$$

For instance, an apartment building manager considering the implementation of water sub-meters would want to know both the costs of the water sub-meters as well as the financial benefits. (Water sub-meters are water meters place on individual apartments where water consumption is measured and billed by the landlord directly, who then in turn pays for the entire building's water through the property's main water meter). A salesman might indicate that a $25,000 investment in water meters improves property value by $300,000. Hence, the ROI on the water sub-meters is calculated at

$$ROI = \frac{\$300{,}000 - \$25{,}000}{\$25{,}000} = 11 = 1100\%$$

Given that the ROI on this investment is 1,100%, the apartment manager is likely to make the investment and purchase the water sub-meters.

ROI is calculated for a number of industrial products. ROI may be calculated over the total lifetime of the product, as was done for the water meters, over a 3 year period, for other semi-durable industrial products, or over a single year, for a consumable industrial product.

Exercises

Lakeview Neighborhood

Resident groups in the gentrified Lakeview neighborhood were randomly assigned to one of three groups. Each group was sent a different mail concerning the importance of greenery/garden plantings in controlling urban pollution. The mayor approved three mailings which differed by:

- Stressed the **positive** benefits of greenery/garden plantings.
- Stressed the **negative** impact of not having greenery/garden plantings.
- Did **not mention** benefits or negatives of greenery/garden plantings, but instead provided information on neighborhood events.

Garden experts rated persons in each group on a 10-point scale, with 10 indicating the greatest variety in garden plantings and 1 indicating no variety in plantings.

No mention	Negatives	Positives
6	8	7
1	7	8
3	6	10
2	9	10
7	9	9

Using One-Way ANOVA in Excel, at the 5% significance level, determine the following:

1. Is the p-Value of the F statistic greater than or less than α?
2. Should we reject or accept the null hypothesis?
3. Is the difference in sample means statistically significant?
4. Are the sample groups from different populations?
5. Explain your findings in words addressing the implications of these findings for future mailings.
    a. Does the message impact the variety in garden planting? If so, which has the highest impact and which has the lowest? Quantify your statements with the means. If not, quantify the overall mean of the sample.
    b. Would you suggest future mailings should stress the positive benefits, focus on the negative impact or do neither?

Insurance Sales Training

The personnel manager of a large insurance company wishes to evaluate the effectiveness of four different sales-training programs designed for new employees. A group of 32 recently hired college graduates are randomly assigned to the four programs such that there are eight persons in each program. At the end of the month-long training period, an exam was administered to the respondents.

PROGRAMS			
A	B	C	D
66	72	61	63
74	51	60	61
82	59	57	76
75	62	60	84
73	74	81	58
97	64	55	45
87	78	70	69
88	53	71	70

Using One-Way ANOVA in Excel, at the 5% significance level, determine the following:

1. Is the difference in sample means statistically significant?
2. Discuss the findings in a summary that the personnel manager might send to the vice-president of the company, i.e., in words that non-statisticians will understand but which includes quantification of the relevant averages.

NEXT IS JACKSON: TRANSFER TO ALL COLORS OF THE RAINBOW AT JACKSON.

In a Red, Brown, and Purple CTA train bumper test, three cars from each train line were crashed into a barrier at 5 mph, and the resulting damage was estimated. Crashes were from three angles: head on, slanted, and rear-ended. The results are shown below.

5 mph Collision Damage ($)			
Crash Type	**Red Line**	**Brown Line**	**Purple Line**
Head-On	700	1,700	2,280
	1,400	1,650	1,670
	850	1,630	1,740
Slant	1,430	1,850	2,000
	1,740	1,700	1,510
	1,240	1,650	2,480
Rear-end	700	860	1,650
	1,250	1,550	1,650
	970	1,250	1,240

Using Two-Way ANOVA with replication in Excel, at the 5% significance level, determine the following:

1. Is the mean repair cost affected by crash type and/or vehicle type?
2. Is there an interaction effect?
3. Are the observed effects, if any, large enough to be of practical importance, as opposed to statistical significance?

CONCRETE PAVEMENT MACHINE

A concrete pavement machine costs $350,000 but is anticipated to do the work of 10 workers. Assume that the average concrete paving worker is paid $40 per hour inclusive of all salary, payroll taxes, healthcare, and other benefits.

1. If the average worker works 8 hours a day, 5 days a week, 50 weeks a year, how many hours a year does the worker work? (The average American worker gets 2 weeks off for vacation each year, reducing the working weeks from 52 per year to 50 per year).
2. If each worker costs the firm $40 per hour, how much does each worker cost the firm per year?
3. If the concrete pavement machine reduces the need for 10 workers, how much does one machine save a customer each year?
4. What is the one-year ROI on a concrete pavement machine?

GROUPON

A restaurant has been asked by Groupon to run an online group coupon promotion. Groupon incurs a cost to the restaurant through the redemption of coupons as denoted by total campaign costs. Groupon delivers a gain to the restaurant through the sale of coupons and through increased restaurant profits. After the promotion is completed, the restaurant calculates that the following costs and profits were incurred.

Total Campaign Costs	**$151,000**
Coupon Sales Revenue	$ 50,000
Restaurant Profits Attributed to Groupon	$119,000
Total Campaign Profits	**$169,000**

What was the ROI of this Groupon promotion for the restaurant?

CHAPTER 8: CHI-SQUARE AND NOMINAL DATA

While t-tests and F-tests are useful for comparing numerical data between two or more samples respectively, they are inappropriate for comparing nominal data between samples. **Pearson's Chi-square test is used to compare samples when nominal data has been collected.** (Chi is pronounced like "why" with a hard K, and is often spelled with the Greek letter χ, as in χ^2).

As with other statistical tests of sample similarity, the analyst can rest easy knowing that the Chi-square test also results in a p-value that can be compared to a chosen significance level in a hypothesis test.

In this chapter, students will learn:

- How to conduct a chi-square test on nominal categorical data.
- How to identify the degrees of freedom associated with a chi-square test.
- How to calculate discount rates and net present value.

CHI-SQUARE

Nominal data, as introduced in Chapter 1, is data which categorizes items into groups. These groups might indicate market segments, purchase patterns, or other marketing variables.

For instance, researchers may collect the results of a simple binary categorical question, like "does the customer purchase or not?" Or, they may be examining age groups in relationship to purchase behavior. For comparing samples when categorical data has been collected, we will rely on Pearson's Chi-square (denoted by χ^2).

Chi-square tests are common with many forms of research. Importantly, commonly performed A/B tests will rely on chi-square statistics to determine if a change of a specific marketing factor resulted in a change in consumer behavior.

Fundamentally, a chi-square test compares the distribution of the two independent samples against a distribution that would have been expected if the two samples are from the same population. If the probability that two distributions arise from the same population is sufficiently high, then there is no statistical difference between the sample distributions, and both samples should be described with the overall distribution. If the probability that two distributions arise from the same population is sufficiently low, then there is a statistical difference between the sample distributions, and each sample should be described by their independent distributions.

For instance, a marketing manager is testing the efficiency of his webpage layout and considers moving the image of the product from the left to the right of the page. That marketing manager can perform an A/B test to determine which layout is most efficient at converting page viewers into customers. Randomly, web visitors can be sorted into two groups, Group A and Group B. Group A visitors see the image on the left, and Group B visitors see the image on the right. After a number of visitors have been to the page, the marketing manager measures their purchase frequency. Suppose the following data is collected in such an A/B test:

Group	Group Size (N)	Purchase Frequency
A	1327	2.79%
B	1433	1.63%

While it looks like Group B had a lower purchase frequency than Group A, we can't yet be sure if this difference is due to random sampling error or a real difference in purchase rates.

To determine if the difference is real, the researcher performs a chi-square test.

- First, the researcher arranges the data in the following format.

Observed Data			
		Purchase	No Purchase
Group	A	37	1290
	B	23	1410

- Second, the researcher calculates the sums of each row and column as a process step towards generating an expected distribution if the two samples derived from the same population.

Observed Data				
		Purchase	No Purchase	Sum
Group	A	37	1290	1327
	B	23	1410	1433
	Sum	60	2700	2760

- Then, the analyst asks "what did I expect?" If there were no differences between the two groups (the null hypothesis), what would have been the expected purchase frequencies? To generate the expected frequencies, the analyst calculates the weighted average response for each category given the overall response pattern.
    - If the two groups were the same, then the response percentages would be the same. On average, we would have expected 60 / 2760 visitors to purchase, and 2700 / 2760 visitors not to purchase in both Groups A and B.
    - For specific groups, we calculate the expected response frequencies as the overall response probability times the number of responses from that group.

$$Group's\ Response\ Item\ Frequency$$
$$= Number\ of\ Responses\ from\ the\ Group\ \cdot Overall\ Responses\ Probability$$

-
    - Hence, for Group A Purchase, we would have expected 1327 · (60/2760), or the number of Group A members times the overall probability of purchasing; for Group A No Purchases, we would have expected 1327 · (2700/2760), or the number of Group A members times the overall probability of not purchasing.
    - A similar calculation is conducted from the remaining sample groups and response items.
    - The result is a tabulation of Expected Data, or data which would have been expected to have been observed if the different sample groups arose from the same population.
    - The items within the Expected Data table would be calculated as follows:

Expected Data				
		Purchase	No Purchase	Sum
Group	A	$1327 \cdot \frac{60}{2760}$	$1327 \cdot \frac{2700}{2760}$	1327
	B	$1433 \cdot \frac{60}{2760}$	$1433 \cdot \frac{2700}{2760}$	1433
	Sum	60	2700	2760

- Yielding the following table of Expected Data

Expected Data				
		Purchase	No Purchase	Sum
Group	A	29	1298	1327
	B	31	1402	1433
	Sum	60	2700	2760

The chi-square test compares the observed data with the expected data, where the expected data is that which would have been expected in the absence of any difference between the two samples.

The next step is to actually perform the chi-square test. Similar to other statistical tests of significance, we construct a hypothesis test to determine if the difference between the actual data and the expected data is due to random sample variations or real differences between the groups.

Choosing the 5% significance level ($\alpha = 0.05$), we compare the p-value of a chi-squared test to the significance level. For the above data, we will find the p-value is 0.033. Since the p-value is less than the alpha, we reject the null hypothesis and declare the differences to be real. Group A purchased more frequently than Group B and putting the image on the left was thus more effective than putting the image on the right.

Chi-Square and Excel

To conduct a chi-square test in Excel, we use the CHIQ.TEST(Observed Array, Expected Array) function. Since CHIQ.TEST uses arrays of numbers, the arrays of observed values must be similarly arranged as those for the expected values. One method to similarly arrange the observed and expected values is simply to stack the two sets of data with the observed data above the expected.

Continuing with the example of page layouts, the analyst might arrange the data as such.

	A	B	C	D	E
1	**Observed**				
2			Purchase	No Purchase	Sum
3	Group	A	37	1290	1327
4		B	23	1410	1433
5		Sum	60	2700	2760
6					
7	**Expected**				
8			Purchase	No Purchase	Sum
9	Group	A	29	1298	1327
10		B	31	1402	1433
11		Sum	60	2700	2760
12					
13	Chi Test	p- Value	0.033		

Where, in cell C13, the analyst input the formula "=CHISQ.TEST(C3:D4,C9:D10)". The CHIQ.TEST() function returns the p-value of the chi-squared test for similarity.

Degrees of Freedom and Chi-Square

A critical issue in a chi-square test is the number of degrees of freedom, often denoted by df. In the above exploration of chi-squared, there was only 1 degree of freedom. Things either did or didn't happen. There was only one dimension of possible variation.

In general, the degrees of freedom is the number of values in the final calculation of a statistic that are free to vary. For a chi-square test, the degrees of freedom can be calculated as the product of the number of samples less one and the number of categories less one.

$$Degrees\ of\ Freedom = (Number\ of\ Samples - 1) \cdot (Number\ of\ Categories - 1)$$

In an excel spreadsheet, we can state this as:

$$Degrees\ of\ Freedom = (Number\ of\ Rows - 1) \cdot (Number\ of\ Colmns - 1)$$

We also mentioned the degrees of freedom when calculating the sample standard deviation. When calculating the mean of a sample, the degrees of freedom are the number of items in the sample. When calculating the sample standard deviation, the degrees of freedom are the number of items in the sample less one. Why "less one"? We use one degree of freedom in calculating the sample mean, and the sample mean is used in calculating the standard deviation, hence the degrees of freedom is reduced by one.

Where might we find more than 1 degree of freedom with categorical data? When more than two types of outcomes are possible. In general, the degrees of freedom in categorical data is equal to the number of categories less one, times the number of samples less one.

For instance, let us consider data on the demographics of different locations and the purchasing of Mexican fast food. Suppose a researcher is comparing two different neighborhoods for a taco and burrito restaurant, the neighborhood at his current location and the neighborhood of a proposed new restaurant location. He believes that age is the main determinant of eating fast Mexican food and therefore records the age profile of people passing by the two store locations. Repeating the same approach as before, the researcher uncovers the following observed data concerning the age demographics near the current location and the proposed new location, calculates the expected profiles if the two neighborhoods were similar, and then identifies the p-value of a chi-square test.

	A	B	C	D	E
1	**Observed**				
2			Current Location	Proposed Location	Sum
3	Age Group	0-19	57	48	105
4		20-39	95	98	193
5		40-65	81	77	158
6		66+	38	42	80
7		Sum	271	262	536
8					
9	**Expected**				
10			Current Location	Proposed Location	Sum
11	Age Group	0-19	53	52	105
12		20-39	98	95	193
13		40-65	80	78	158
14		66+	40	40	80
15		Sum	271	265	536
16					
17	Chi Test	p- Value	0.79		

Because the p-value of the chi-square test of 79% is greater than the significance level of 5%, the researcher may conclude that the two locations have similar age profiles, and hence the taco and burrito restaurant may do well in the new location.

In this data set on age demographics and locations, the degrees of freedom is three.

- The number of locations under consideration is 2, thus the number of samples is 2.
- The number of age groups under consideration is 4, thus the number of categories is 4.
- The number of degrees of is equal to the number of categories less one, times the number of samples less one, or

$$df = (2-1)\cdot(4-1) = 1\cdot 3 = 3$$

Marketing Metrics: Discount Factors and Net Present Value

Return on Investment (ROI), as discussed in Chapter 7, provides an unbiased approach to demonstrating the financial impact of a purchase decision for many types of products or services. ROI is calculated as the ratio of the difference in the value of the benefits and the cost of the product, to the cost of the product. Often, however, the benefits are not immediate. Instead, the benefits accrue over time.

When the benefits accrue over time, the value of those benefits needs to be discounted to account for the time value of money. The concept of the time value of money refers to the fact that a dollar today is worth more than a dollar tomorrow. For instance, consider which you would prefer, to have $10,000 today or receive $10,000 a year from today? Most people would prefer to have the cash today than to have it in a year. The time value of money refers to the decreased value of cash when its receipt is deferred in comparison to the present value of cash in one's hand.

Discount Factors

To calculate the present value of cash, let us work backwards from the starting consideration that every dollar received today could be invested or otherwise put in the bank to collect interest. Let the annual interest rate for an investment be denoted by r. If V_0 dollars were invested in an account earning an interest rate of r, one year from today those V_0 dollars would have accrued interest equal to $V_0 \cdot r$, and the account would hold the original V_0 dollars plus the accrued interest of $V_0 \cdot r$. Hence, the value a year V_1, from now of investing V_0 dollars today is given by

$$V_1 = V_0 + V_0 \cdot r = V_0(1+r)$$

If we divide both sides by (1+r), we can rewrite the above equation after some rearranging as

$$V_0 = \frac{V_1}{(1+r)}$$

V_1 in the numerator on the right hand side of the equation is the value to be received a year from now. V_0 on the left hand side of the equation is the present value of V_1. The denominator of the right hand side of the equation contains the discount rate used to discount V_1 back to present value. (When working backwards in time, the interest rate is referred to as a discount rate even though it is the same number).

Analysts will often have to discount values back to the present value, and as such will often simply calculate discount factors. Discount factors can be multiplied by the future value to determine their present value. Using this definition with the above equation, we find that the discount factor for a one year investment is

$$d_1 = \frac{1}{(1+r)}$$

where we have used the subscript of 1 to denote a one year discount factor.

With the discount factor calculated as defined, we can identify the present value V_0 of the future receipt of V_1 to be

$$V_0 = V_1 \cdot d_1$$

The discount factor for longer periods of time would be similarly identified. For instance, for calculating the present value of a financial benefit which arrives in two years, the discount factor would be

$$d_2 = \frac{1}{(1+r)} \cdot \frac{1}{(1+r)} = \frac{1}{(1+r)^2}$$

and similarly for longer periods of time. In general, the discount factor for n years is given by

$$d_n = \frac{1}{(1+r)^n}$$

With the above definitions, we can calculate the present value as a future benefit and are prepared to uncover net present values.

Net Present Value

The net present value (NPV) of a stream of financial benefits is equal to the total present value of those benefits less the cost to acquire them.

For example, consider a business customer evaluating the purchase of a $25,000 piece of machinery that lasts for five years. The customer anticipates that the machine will begin to save her firm $10,000 each year at the end of the year. Because of the time value of money, she knows she must discount those future savings. Let the discount rate applicable to the firm be 15%, representing the borrowing costs of the firm.

She then creates the following table in Excel and calculates the Net Present Value as well as the ROI.

- Rows 1 and 2 contain the interest rate used by the firm and the cost of the machine respectively.
- Starting in row 4, column A defines the year under consideration and column B defines the savings generated by the machinery for each year under consideration. Column A and B are inputs into the calculation of the net present value.
- In column C, she calculates the discount factor using the equation stated above.
  - Cell C5, the first year's discount factor, was calculated as =1/(1+B1).
  - Cell C6, the second year's discount factor, was calculated as either =1/(1+B1)^2 or =C5/(1+B$1). Both formulae yield the same result.
  - Cell C7 through C9, the third through fifth years' discount factor, was similarly calculated as =C6/(1+B$1).
- In Row 11, she calculates the total present value of all future savings to be $33,521.
- In Row 12, she calculates the net present value (NPV) of the purchase as the difference between total present value of all future savings and the cost to purchase, or $33,521 - $25,000 = $8,522.
- Finally, she calculates the ROI as the ratio of the net present value to the cost to purchase, or $8,522 / $25,000 = 34%.

In this example, the net present value is positive, and the return on investment is significant. As such, she is likely to recommend purchasing the machinery.

	A		B	C	D	E
1	Interest Rate		15%			
2	Cost		$25,000			
3						
4	Year		Future Value	Discount Factor	Present Value	
5		1	$10,000	0.870	$ 8,696	
6		2	$10,000	0.756	$ 7,561	
7		3	$10,000	0.658	$ 6,575	
8		4	$10,000	0.572	$ 5,718	
9		5	$10,000	0.497	$ 4,972	
10						
11	Total Present Value				$ 33,521	
12	NPV				$ 8,522	
13	ROI				34%	
14						

Sales and marketing people who can demonstrate a positive net present value and significant ROI associated with their product or service are far more likely to find a market willing to purchase than those who simply highlight the cost of the product.

EXERCISES

COLD CALLS

A salesperson is testing the right approach to manage a personal assistant in an effort to speak with an executive. This salesperson determines to separate his outbound sales calls into two groups. With group A, the salesperson chooses to reveal no information to the personal assistant other than their desire to speak with the executive. With group B, the salesperson chooses to reveal the nature of their call to the personal assistant and ask for assistance in determining the best approach to contact the executive. After two weeks of phone calls, the salesperson gathers the following data:

Observed Data			
		Talked with Executive	Did Not Talk with Executive
Group	A	110	343
	B	182	428

1. Within each sample group A and B, what are the response probabilities of "Talk with Executive" and "Did Not Talk with Executive"?
2. Create a 100% stacked bar chart of the data that has both groups on one plot.
3. Calculate the sum across both rows and both columns.
4. Calculate the expected frequencies if there were no differences between the two groups.
5. Calculate the p-value of a chi-square test on the data and interpret the results at the 5% significance level.
    a. Are the two groups different at a statistically significant level?
    b. Should we reject or accept the null hypothesis?
    c. Does the way the salesperson interacts with an administrative assistant make a difference? If so, make a quantitative statement regarding the average of each or both groups. If not, make a quantitative statement regarding the overall average of the combined two groups.
6. How many degrees of freedom does this data set have?

PBS Viewers

A researcher hypothesized that persons who were college educated would watch PBS programs more frequently than those who had high school degrees. These data are shown below.

Observed Data			
		College Degrees	High School Degrees
Watch PBS	Often	37	10
	Occasionally	51	32
	Never	8	54

1. Create a cross tab of the data and plot the data in a 100% stacked bar chart.
2. Calculate the p-value of a chi-square test on the data and interpret the results at the 5% significance level.
    a. Are the differences statistically significant?
    b. Is the researcher's hypothesis correct?
    c. How would you explain these results to a senior executive?
3. How many degrees of freedom does this data set have?

Starbucks

A researcher hypothesized that older persons were not as good customers for Starbucks as were younger persons. These data are shown below.

	Respondents	
Purchase Frequency	Under 25	25 years old or older
Once a week or more	40	7
Once a month to once a week	51	32
Once a year or less	8	54

1. Create a cross tab of the data and plot the data in a 100% stacked bar chart.
2. Calculate the p-value of a chi-square test on the data and interpret the results at the 5% significance level.
    a. Are the differences statistically significant?
    b. Is the researcher's hypothesis correct?
    c. Does one of the market segments purchase more frequently than the other?
    d. How would you explain these results to a senior executive?
3. How many degrees of freedom does this data set have?

Morning or Evening Selling is Best?

A salesperson is testing two different approaches to engaging a customer. Prospects in group A were contacted before 10 am in the morning. Prospects in group B were contacted after 3 pm in the afternoon. The success in winning or losing a sale was tracked as "Won" or "Lost".

1. How many wins appear in Group A? In Group B?
2. How many losses are in Group A? Group B?
3. Create a cross tab of the data and plot the data in a 100% stacked bar chart.
4. What is the mean of means between the two samples for the win ratios?
5. Calculate the p-value of a chi-square test on the data and interpret the results at the 5% significance level.
    a. Are the differences statistically significant?
    b. If a meeting had to be scheduled, when should the meeting be held and when would you suggest the salesperson place his calls?
6. How many degrees of freedom does this data set have?

Group A				Group B			
Lost	Lost	Lost	Lost	Won	Lost	Lost	Lost
Lost	Lost	Lost	Won	Lost	Lost	Lost	Lost
Won	Lost	Lost	Won	Lost	Lost	Lost	Lost
Lost	Lost	Lost	Lost	Lost	Lost	Lost	Lost
Lost	Lost	Lost	Lost	Lost	Lost	Lost	Lost
Lost	Won	Lost	Lost	Lost	Lost	Lost	Lost
Lost	Lost	Lost	Lost	Lost	Lost	Lost	Lost
Lost	Won	Lost	Lost	Lost	Lost	Lost	Won
Lost	Won	Lost	Lost	Lost	Lost	Lost	Lost
Lost	Lost	Lost	Lost	Lost	Lost	Won	Lost
Won	Lost	Won	Lost	Lost	Lost	Lost	Lost
Lost	Lost	Lost	Won	Won	Lost	Lost	Lost
Lost	Lost	Won	Lost	Lost	Lost	Lost	Lost
Lost	Lost	Lost	Lost	Lost	Lost	Lost	Lost
Lost	Lost	Lost	Lost	Lost	Lost	Lost	Lost
Lost	Won	Lost	Lost	Lost	Lost	Lost	Lost
Won	Lost	Lost	Won	Lost	Lost	Lost	Lost
Lost	Won	Lost	Lost	Lost	Won	Lost	Lost
Lost	Won	Lost	Won	Lost	Lost	Lost	Lost
Lost	Lost	Won	Lost	Lost	Lost	Lost	Won
Won	Lost	Lost	Lost	Lost	Lost	Lost	Lost
Lost	Lost	Lost	Lost	Lost	Won	Lost	Lost
Lost	Lost	Lost	Lost	Lost	Lost	Lost	Lost
Won	Lost	Lost	Won	Lost	Lost	Lost	Won

MALE OR FEMALE PHOTOS

An internet marketer is testing two different approaches to engaging a customer. Website visitors in group M were shown the product held by a male model. Website visitors in group F were shown the product held by a female model. The success in sales was tracked as "Bought" or "Left".

1. How many wins appear in Group M? In Group F?
2. How many losses are in Group M? Group F?
3. Create a cross tab of the data and plot the data in a 100% stacked bar chart.
4. Calculate the p-value of a chi-square test on the data and interpret the results at the 5% significance level.
   a. Are the differences statistically significant?
   b. If an executive had to choose a picture and had a desire for a female photo, would you agree, disagree, or be indifferent to his choice?
5. How many degrees of freedom does this data set have?

Group M				Group F			
Left	Left	Left	Left	Bought	Left	Left	Left
Left	Left	Left	Bought	Left	Left	Left	Left
Left	Left	Left	Bought	Left	Left	Left	Left
Left	Left	Left	Left	Left	Left	Left	Left
Left	Left	Left	Left	Left	Left	Left	Left
Left	Bought	Left	Left	Left	Left	Left	Left
Left	Left	Left	Left	Left	Left	Left	Left
Left	Left	Left	Left	Left	Left	Left	Bought
Left	Left	Left	Left	Left	Left	Left	Left
Left	Left	Left	Left	Left	Left	Bought	Left
Left	Left	Left	Left	Left	Left	Left	Left
Left	Left	Left	Left	Bought	Left	Left	Left
Left	Left	Left	Left	Left	Left	Left	Left
Left	Left	Left	Left	Left	Left	Left	Left
Left	Left	Left	Left	Left	Left	Left	Left
Left	Left	Left	Left	Left	Left	Left	Left
Bought	Left	Left	Left	Left	Left	Left	Left
Left	Bought	Left	Left	Left	Left	Left	Left

Tradeshow Advertising

A marketer is considering sponsoring a tradeshow in exchange for an exhibit in the exhibit hall. While the marketer expects no immediate sales from the tradeshow, she does expect a sale to be generated from the tradeshow within a year. Suppose the profits from that sale are expected to be $30,000 and the cost of the tradeshow sponsorship plus attendance is expected to be $10,000. Given a 25% discount rate, calculate the following:

1. The one year discount factor.
2. The present value of the future profits from the sale.
3. The net present value (NPV) of sponsoring the trade show.
4. The return on investment (ROI) for sponsoring the trade show.
5. Should the marketing manager sponsor the tradeshow?

Monument Etching Machine

A manufacturer of etching machines used in the funeral monument industry desires to communicate the value of his machine. The machine costs $17,900 and is anticipated to last 5 years. Each year, the machine is estimated to save a funeral monument firm $30,000. Given a 15% discount rate, calculate the following:

1. The discount factor for years one through five.
2. The present value of the future savings for each year.
3. The net present value (NPV) of the monument etching machine.
4. The return on investment (ROI) for a customer of the machine.
5. Would a customer find this to be a compelling reason to purchase?

Employee Training

A training firm desires to communicate the value of their training services. Their training course costs $5,000 per employee. The average employee stays with a firm for 3 years. The benefits of the training are not anticipated to arrive evenly. In the first year, the training is anticipated to improve the productivity of the employee by $2,500. In the second year, the training is anticipated to improve the productivity of the employee by $5,000. In the third year, the training is anticipated to improve the productivity of the employee by $10,000. Given a 20% discount rate, calculate the following:

1. The discount factor for years one through three.
2. The present value of the future productivity gains for each year.
3. The net present value (NPV) of the training.
4. The return on investment (ROI) for training.
5. Would a customer find this a compelling reason to purchase?

CHAPTER 9: REGRESSION ANALYSIS

We have examined relationships between distinct samples using both categorical data (Chi-square tests) and numerical data (t-tests and F-tests). The last type of relationship we will explore is not between distinct samples but rather influences that can numerically vary continuously on outcomes that can also numerically vary continuously. Regression analysis examines correlations between numerical data.

For instance, a marketer might suspect that sales volume is related to price. Both price and sales volume are ratio data. t-tests, F-tests, and Chi-square tests are not appropriate for evaluating this relationship. Instead, the analyst will need to regress sales volume (the dependent variable) on price (the independent variable) to explore the relationship between price and volume.

Once a relationship between two numerical variables has been determined, the research can then make predictions. For instance, if the marketer determines that a $1 price increase leads to a 25 unit sales volume decrease, the marketer can predict that a $4 price decrease would yield a 100 unit sales volume increase.

In this chapter, students will learn:

- Relationships between independent and dependent variables for making predictions.
- How to create scatter plots to explore relationships between variables.
- The use of correlation coefficients for clarifying the relationships between variables.
- The use of Pearson's R for quantifying the strength of a relationship between variables.
- How to conduct linear and multivariate regression to clarify the relationships between variables and make predictions.
- How to interpret the output of an Excel regression analysis.
- A method to explore the effect of a price change on sales volume.

VARIABLES

Are two items related? Does age indicate interest in healthcare? Does distance from a coast indicate demand for seafood? Is exposure to advertising correlated with willingness to purchase a product? Do higher prices lead to fewer sales? Is product quality related to demand?

Strategic questions like these require looking at how variables are related. **Correlation and regression analysis examines how variations in one variable are related to those in another.**

In conducting a correlation or regression analysis, we divide variables into two key categories.

Independent variables are variables that are free to vary on their own. Often, independent variables are marketing input which can be manipulated. Dependent variables are variables in which variations are, in some way, driven by variations of independent variables.

Hence, for the above questions, we could categorize the variables in the questions as either independent or dependent.

Independent	Dependent
Age	Healthcare Interest
Distance from Coast	Seafood Demand
Advertising Exposure	Willingness to Purchase
Price	Quantity Sold
Product Quality	Demand

Scatter Plots

To examine if two variables are related, we often make scatter plots. Scatter plots graph the data with one variable, usually the independent variable, on the horizontal axis and the other variable, usually the dependent variable, on the vertical axis.

For instance, if a researcher wanted to know if proximity to a coast is related to fish eating, they may survey several individuals across the nation and determine the percentage who eat fish once a week. If the following data was collected, they could graph the data with a scatter plot placing the percentage of individuals that eat fish weekly against coastal distance.

From looking at the scatter plot of the percentage of frequent fish eaters against coastal distance, it appears as though coastal proximity does indeed influence fish eating. Data points corresponding to people far from the coastline also correspond to a lower percentage of weekly fish eaters. In contrast, data points corresponding to people near the coastline also correspond to a higher percentage of weekly fish eaters. Hence, the research might suspect that distance from the coast is negatively correlated with interest in seafood.

Scatter plots can be created with Excel in a very similar manner as any other plot.

As with many issues in marketing, graphs help in interpreting and presenting data. However, to determine if the relationships are significant, we have to undertake a statistical analysis.

Distance from Coast (miles)	Percentage of Weekly Fish Eaters
661	68
748	50
1600	63
60	99
1087	8
1879	2
87	95
635	84
226	90
1418	86
556	74
696	85
2905	27
365	82
1470	65
269	90
400	89
69	94
365	90
52	98

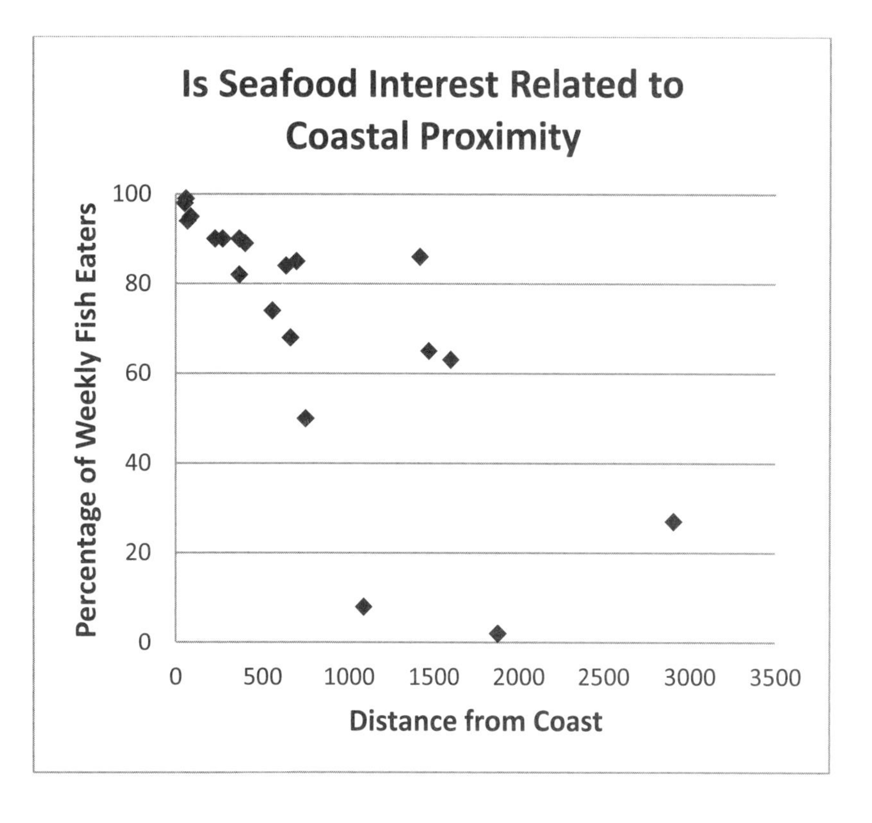

CORRELATION

Correlation quantifies the relationship between two numerical variables, determining if they are related.

Items are positively correlated if the increase in one variable is associated with an increase in another variable. One could also have said that items are positively correlated if the decrease in one variable is associated with a decrease in another variable. For example, educational attainment (independent variable) might be positively correlated with lifetime earnings (dependent variable). Higher educational attainment is hence associated with higher lifetime earnings, and lower educational attainment is associated with lower lifetime earnings.

Items are negative correlated if the increase in one variable is associated with a decrease in another variable. One could also have said that items are negatively correlated if the decrease in one variable is associated with an increase in another variable. For example, higher levels of income may be associated with lower levels of interest in salt pork and lower levels of income might be associated with a higher interest in salt pork.

Pearson's R

To calculate the level of correlation, we identify Pearson's R, also known as the coefficient of correlation.

$$R = \frac{\sum(X_i - \bar{X}) \cdot (Y_i - \bar{Y})}{\sqrt{\sum(X_i - \bar{X})^2 \cdot \sum(Y_i - \bar{Y})^2}}$$

Pearson's R varies between -1 and 1. When Pearson's R is less than zero, the correlation would be negative. When Pearson's R is greater than zero, the correlation would be positive.

The absolute value of Pearson's R also tells us something about the strength of the correlation. As the absolute value of Pearson's R approaches one, the correlation becomes stronger. As the absolute value of Pearson's R approaches zero, the correlation becomes weaker.

In interpreting Pearson's R with respect to correlation strength, there is no universal rule. However, we can offer the following guidelines.

Absolute Value of Pearson's R	Strength of Correlation
1 > R > 0.8	Very Strong
0.8 > R > 0.6	Strong
0.6 > R > 0.4	Moderate
0.4 > R > 0.2	Weak
0.2 > R	Very Weak

R-Squared

If we square Pearson's R, we get a value named R-Squared. R-Squared is a powerful metric. It describes the percentage of variation in the dependent variable that is determined by the independent variable.

Correlation Is Not Causation

Correlations have tremendous predictive value, yet correlation is not causation, not all correlations are meaningful, and one should be careful about extrapolating correlations to ranges outside of those measured.

One major challenge of assuming correlation implies causation is the presence of third variables. For instance, the presence of grey hair may be correlated with wrinkled skin, but I doubt anyone would state that grey hair causes wrinkled skin. A third variable, AGE, is more likely to cause grey hair and wrinkled skin.

Another challenge to correlations is the presence of outliers. If most of the data lies in a small area of the scatter plot, but a few data points lie a very large distance from the others, those few data points may drive a spurious correlation calculation. Outliers can both increase and decrease Pearson's R. When outliers are identified, researchers are often advised to either remove it from the dataset, or collect more data around that point.

A third challenge to correlation is the inability to reliably extrapolate a correlation found in one area to all ranges of possible variation. For instance, between the ages of 0 and 16, one might find a clear correlation between age and height. However, extrapolating these results to an entire lifetime would lead one to conclude that 90 year old people must be towering giants. Observation alone refutes this conclusion.

Regression

If two variables are correlated, we can use one variable to predict the value of another variable. Regression enables the researcher to specify the relationship between variables for prediction.

Slope and Intercept

We can draw regression line through correlated data on a scatter plot. Like all lines, the regression line is defined by a slope and an intercept.

$$Y = mX + b$$

Regression enables us to identify the specific slope and intercept that describes the relationship between the independent variable X and the dependent variable Y.

Once the regression line is identified, the relationship can be used to predict the value of the dependent variable Y when given a value for the independent variable X.

In regression analysis, the slope is found through the expression

$$m = \frac{N \sum X_i Y_i - \sum X_i \cdot \sum Y_i}{N \sum X_i^2 - (\sum X_i)^2}$$

Notice, unlike the expression for Pearson's R, the slope of the regression line is not symmetric in X and Y. Only the independent variable appears in the denominator.

Therefore, prior to conducting a regression analysis, the researcher must determine which variable is the dependent variable and which variable is the independent variable. Moreover, a regression equation that defines Y as dependent on X cannot be rearranged to predict X with a given Y. Regression lines only go one way.

Once the slope is uncovered, we can uncover the intercept. The intercept is found through the expression

$$b = \bar{Y} - m\bar{X}$$

Simply using the average X and Y along with the uncovered regression slope will reveal the intercept.

Interpreting Excel Regression Output

While calculating Pearson's R, R-Squared, Regression Slopes, and Regression Intercepts with the above equations is a straightforward exercise, Excel will perform the task for us.

Using "Tools" / "Data Analysis" / "Regression" in Excel, we will find a menu of options for conducting the regression analysis. Put the independent data in a column next to the dependent data. For input Y range, select the dependent data. For the input X range, select the independent data. Choose where you want the output, and Excel will do the rest of the mathematical analysis for you.

Below is some of the output one would get from Excel after regressing the percentage of weekly fish eaters on the distance from coast. To interpret this output, pay close attention to a few key metrics.

SUMMARY OUTPUT

Regression Statistics	
Multiple R	0.742889
R Square	0.551884
Adjusted R Square	0.526989
Standard Error	19.92248
Observations	20

ANOVA

	Df	*SS*	*MS*	*F*	*Significance F*
Regression	1	8798.659	8798.659	22.16817	0.000175
Residual	18	7144.291	396.905		
Total	19	15942.95			

	Coefficients	*Standard Error*	*t Stat*	*P-value*
Intercept	94.39248	6.524219	14.4680	2.36E-11
X Variable 1	-0.02887	0.006131	-4.7083	0.00017

MULTIPLE R

Multiple R is the absolute value of Pearson's R. In this case, the absolute value of Pearson's R is 0.74, implying that the correlation is strong.

R-SQUARED

R-Squared is 0.55, implying that this dataset indicates that distance from the coast predicts 55% of the difference in the percentage of individuals who eat fish weekly.

Coefficients

The intercept coefficient is 94.4 and the X variable 1 coefficient is -0.029. This implies that if distance from the coast could be used to predict the percentage of individuals who eat fish weekly, the slope would be -0.029 percentage points / mile and the intercept would be 94.4 percentage points.

$$\text{Percentage of Weekly Fish Eaters} = -0.029 \cdot \text{Distance From Coast (mi)} + 94.40$$

T-Stat

The t-stat for the intercept is 14.47 and that for the slope is -4.708. In many ways, the t-stat is like a Z score: it tells you how many standard deviations a measurement is away from zero. Since both of the t-stats are greater than 1.96 in absolute value, corresponding to the 95% confidence interval, we can state with confidence that our intercept and slope are statistically significant non-zero numbers.

P-Value

The p-value for the intercept is 0.0000 and that for the slope is 0.0002. As with p-values from other tests, we can compare these p-values against α of the chosen significance level. Since both of the p-values are less than $\alpha = 0.05$, we can reject the null hypothesis and conclude that the intercept and slope differ from zero in a statistically significant manner.

Multivariate Regression

While the above formulae for regression analysis work for examining the relationship between a dependent variable and one independent variable, other much more demanding expressions are required to analyze the relationship between a dependent variable and two or more independent variables.

Fortunately, Excel will do the math for us, and we don't have to worry about these formulae.

When we examine the relationship between a dependent variable and multiple independent variables, we are doing a multivariate regression analysis.

In Excel, the procedure and interpretation of the output is much the same for multivariate regression analysis as it is for bivariate regression analysis. The one key difference arises in selecting the Input X range. For simple bivariate regression analysis, the Input X range is a single column of the single independent variable under investigation. For multivariate regression analysis, the Input X range is a collection of columns, one for each independent variable under investigation. Consequently, the output of the regression analysis will yield coefficients, t-stats, and p-values for each X variable. Taking care to recall the order of the variables, we can interpret the results of a multivariate regression in the same manner as we did with a simple linear regression.

MARKETING METRICS: PRICE AND DEMAND

PRICE TO QUANTITY SOLD

In most cases, marketers are aware that higher prices lead to lower sales volumes, and lower prices lead to higher sales volumes. We can examine the relationship of price to quantity sold in a manner much like that for considering the correlation of any other two variables.

In evaluating customer demand, price is the independent variable and quantity sold is the dependent variable. While we usually plot the dependent variable on the vertical axis and the independent variable on the horizontal axis, economic tradition dictates otherwise. Hence, a plot of units sold to price for a firm might look like the following.

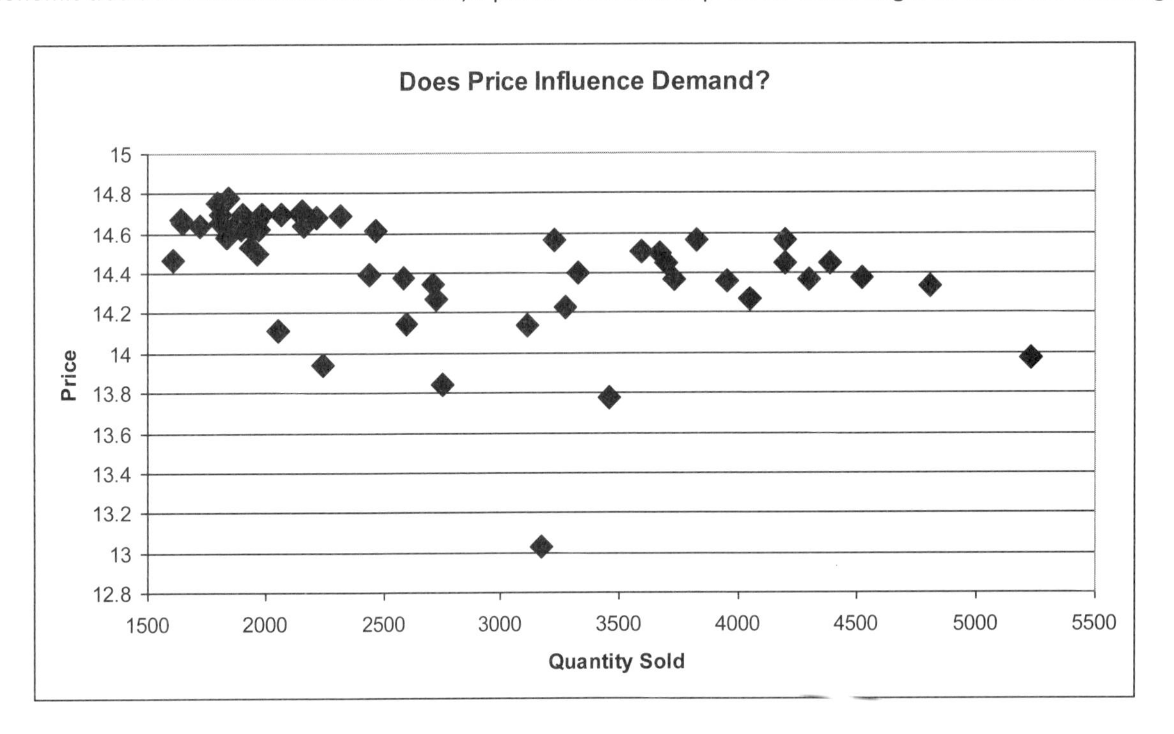

When conducting a regression analysis of this data, researchers should be careful to ensure to input price as the independent X Range and quantity sold as the dependent Y Range.

As this image of quantity sold to price indicates, the correlation is week. Moreover, the data point at $13 and 3200 units appears to be an outlier.

Removing this data point, one finds:

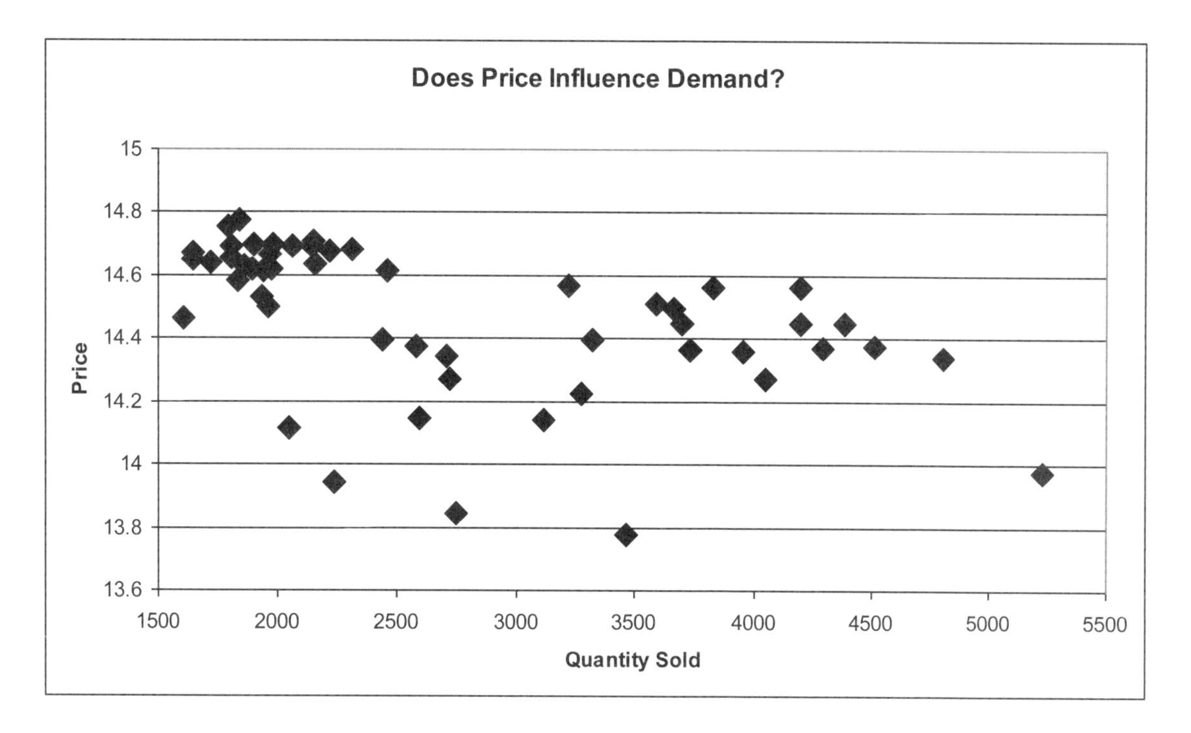

From this second graph, it appears as if there is a negative correlation between prices and sales volumes. The next step is to complete the regression analysis. Taking care to ensure that price is used as the independent variable in X range and quantity sold is used as the dependent variable in the Y range, we get the following output.

SUMMARY OUTPUT

Regression Statistics	
Multiple R	0.476071
R Square	0.226643
Adjusted R Square	0.211176
Standard Error	880.0906
Observations	52

ANOVA

	df	*SS*	*MS*	*F*	*Significance F*
Regression	1	11349789	11349789	14.65322	0.000361
Residual	50	38727972	774559.4		
Total	51	50077761			

	Coefficients	*Standard Error*	*t Stat*	*P-value*	*Lower 95%*
Intercept	31316.8	7462.146	4.196755	0.000111	16328.64
X Variable 1	-1974.25	515.7445	-3.82795	0.000361	-3010.15

From the regression analysis, both the t stat and the p-value indicate that the slope and intercept differ from zero in a statistically significant manner. However, with an R Squared of only 23%, we can assume that much of the variation in quantity sold is driven by factors other than price for this dataset.

EXERCISES

REAL ESTATE VALUES

Below are the appraised values and the sales prices for five different real estate properties.

1. Create a scatter plot of Sales Price (vertical axis) against Appraised Value (horizontal axis).
2. What is Pearson's R and the strength of the regression?
3. What is R-Squared and what percentage of the variation in Sales Price is explained by the variation in Appraised Value?
4. Is the slope of the regression statistically significant? Is the intercept statistically significant?
5. Create an equation for predicting the Sale Price given an Appraised Value.
    a. If the Appraised Value is 2.5, what would the expected Sales Price be?
    b. If the Appraised Value is 4.5, what would the expected Sales Price be?
    c. If the Appraised Value is 8, what would the expected Sales Price be?
6. Is there a causal relationship between the two variables? Why do you say that?

Appraised Value*	Sale Price*
2	2
3	5
4	7
5	10
6	11

*In $100,000's

TIVOLI PROMOTION BUDGET MANAGEMENT

A sample of 34 stores in a supermarket chain is selected for a test-market study of Tivoli premium chocolate bars. All the stores selected have approximately the same monthly sales volume. Two independent variables are considered here – the price of a Tivoli premium chocolate bars, as measured in cents (X1), and the monthly budget for in-store promotional expenditures, measured in dollars (X2). The dependent variable Y is the number of Tivoli premium chocolate bars sold in a month.

1. Create a scatter plot of
    a. Sales (vertical axis) against Price (horizontal axis).
    b. Sales (vertical axis) against Promotion (horizontal axis).
2. Regress Sales
    a. Price alone
    b. Promotion alone
    c. Price and Promotion simultaneously
3. Which regression yields the best results?
    a. In which regressions is the intercept statistically different from zero?
    b. Is Sales statistically correlated to Price? To Promotion? To both?
    c. Which regression yields the highest R-Squared?
4. Create an equation for predicting the sales given appropriate input.
    a. If the Price is 69 and the Promotion is 300, what would the expected Sales be?
    b. If the Price is 89 and the Promotion is 500, what would the expected Sales be?
5. Is there a causal relationship between the dependent variable and the independent variables? Why do you say that?

Store	Sales	Price	Promotion
1	4,141	59	200
2	3,842	59	200
3	3,056	59	200
4	3,519	59	200
5	4,226	59	400
6	4,630	59	400
7	3,507	59	400
8	3,754	59	400
9	5000	59	600
10	5,120	59	600
11	4,011	59	600
12	5,015	59	600
13	1,916	79	200
14	675	79	200
15	3,636	79	200
16	3,224	79	200
17	2,295	79	400
18	2,730	79	400
19	2,618	79	400
20	4,421	79	400
21	4,113	79	600
22	3,746	79	600
23	3,532	79	600
24	3,825	79	600
25	1,096	99	200
26	761	99	200
27	2,088	99	200
28	820	99	200
29	2,114	99	400
30	1,882	99	400
31	2,159	99	400
32	1,602	99	400
33	3,354	99	600
34	2,927	99	600

GAMBLING, AND CRIME RATES, AND CORRELATION! OH, MY!

Legalized gambling is available on the Bahamas' cruises. The manager of the cruise wants to know the correlation between the number of casino games per boat and the yearly crime rate. The records for the past 10 years are examined, and the results are shown in the table below. Plot the data and conduct an appropriate regression analysis and interpret the results. Does the number of casino games influence the crime rate?

Year	Number of Casino Games	Crime Rate (per 1,000 pop.)
2000	15	1.45
2001	17	1.73
2002	24	2.43
2003	21	2.30
2004	24	2.61
2005	28	2.83
2006	30	3.51
2007	32	3.27
2008	37	3.64
2009	39	4.35

ALADDIN ISN'T THE ONLY BREAD WINNER

Jasmin is an architect, aside from her job being Princess of Agrabah. She knows that as buildings age, their prices increase linearly. Moreover, Jasmin hypothesizes that the auction price for palaces in Agrabah will increase linearly as the number of bidders increases. Thus, the following first-order model is hypothesized:

$$y = \beta_0 + \beta_1 x_1 + \beta_2 x_2 + \epsilon$$

Where

Y = auction price

x_1 = Age of palace (years)

x_2 = Number of bidders

A sample of 15 auction prices of palaces, along with their age and the number of bidders, is given in the table below.

1. Conduct a multivariate regression analysis of the data.
2. Is Jasmin's hypothesis that the mean auction price of a palace increases as the number of bidders increases accurate? (that is, test $\beta_2 > 0$. Use α = .05).
3. Is Jasmin's hypothesis that the mean auction price of a palace increases as the number of age increases accurate? (that is, test $\beta_1 > 0$. Use α = .05).

Age, x_1	Number of Bidders, x_2	Auction Price (millions), y
97	5	1.6
115	13	1.9
235	17	2.7
256	25	3.0
258	26	3.5
295	29	3.6
299	34	4.2
350	39	4.5
369	40	4.7
386	46	5.2
390	49	5.6
410	58	6.0
427	60	7.8
458	77	7.9
490	92	10.0

Sales by Territory

Assume that a toy manufacturer wishes to forecast sales by sales territory. It is thought that retail sales (because people in the territory are spending), previous sales performance history of the company salesperson and grammar school enrollment (because young children are consumers of toys) are the independent variables that might explain the variation in sales (the single, interval scaled dependent variable). These data are shown below. Use Excel to conduct a multiple regression on these data. Explain your findings.

Sales (Y)	Retail Sales (X1)	Salesperson History (X2)	Grammar School Enrollment (X3)
(000)	(000)	(000)	(000)
222	106	100	23
304	213	100	18
218	201	100	22
501	378	200	20
542	488	100	21
790	509	200	31
523	644	100	17
667	888	200	25
700	941	200	32
869	1066	200	36
444	307	100	30
479	312	200	22

Used Cars

Suppose an auto auction wants to model the relationship between the sale price of used cars in Chicago and the following three independent variables: Age, Basic (0) or Luxury (1), and Gas Mileage (MPG).

Consider the first order model:

$$y = \beta_0 + \beta_1 x_1 + \beta_2 x_2 + \beta_3 x_3 + \epsilon$$

Where

Y = original sale price (dollars)

x_1 = Age in Years

x_2 = Basic (0) Luxury (1)

x_3 = MPG

To fit the model, the appraiser selected a random sample of n = 20 cars from the thousands of cars that were sold in a particular year. The resulting data are given in the table below.

1. Use scatter plots to examine the sample data.
2. Calculate the multivariate regression.
3. What percentage of the variation in price is determined by the regression?
4. To which variable(s) is price correlated?
5. Explain your results to your life partner in words.
6. What price should you expect to earn for a 3 year old basic minivan with 25,000 miles and 15 miles per gallon?

Car #	Sale Price (y)	Years Old (x_1)	Basic or Luxury (x_2)	MPG (x_3)
1	28,900	4	1	22
2	14,500	15	0	10
3	25,500	7	1	22
4	32,000	2	1	27
5	26,500	3	1	32
6	45,000	7	1	16
7	28,000	4	1	30
8	23,000	3	0	24
9	29,000	2	1	22
10	37,500	8	1	27
11	20,500	5	0	29
12	60,000	10	1	15
13	37,000	3	1	10
14	45,500	19	1	18
15	20,900	9	0	28
16	30,000	3	1	26
17	16,000	2	0	28
18	17,000	1	0	25
19	30,000	0	1	21
20	22,500	4	0	22

CHAPTER 10: HYPOTHESIS TESTS

Throughout the second half of this text, we have been using the hypothesis test to make a judgment regarding whether two or more samples are from the same population or from different populations. While the hypothesis test is a useful approach, it isn't flawless.

In this chapter, we will return to the assumptions of the hypothesis test to explore inherent uncertainties with this approach. Specifically, students will learn:

- The differences between Type I errors and Type II errors.
- The challenges of certainty and the need to accept limited uncertainty, also known as humility, in making determinations.

THE NULL HYPOTHESIS

In many sections of this text, we have employed the null hypothesis to determine the statistical significance of a metric.

Student's t Test	**Are the differences in means of numerical data statistically significant?**
ANOVA / F-Test	**Can the variations between numerical data be considered within statistical expectations?**
Pearson's Chi-square	**Are the differences between observed categorical data and the expected categorical data statistically significant?**
Regression Analysis	**Are the regression coefficients statistically different from zero?**

In the t-test, F-Test, and Chi-square tests, the analyst is examining the issue of whether samples are from the same or different populations. In regression analysis, the analyst is determining whether the correlation is statistically different from zero or not. In each case, the analyst uses the null hypothesis to make a determination regarding statistical significance.

However, the approach of comparing things against the null hypothesis is not flawless.

With the null hypothesis, we begin by assuming that the two samples are statistically similar and then attempt to prove this assumption false. (Ayn Rand loves checking assumptions, but would have benefited from checking her own a little more).

The method we use to indicate that the null hypothesis assumption is false is to establish a significance level (alpha) and compare the statistical probability of variations arising due to random sampling error (p-value) against the significance level. If the statistical probability is less than the significance level, we reject the null hypothesis. If not, we accept it.

But each time we accept or reject the null hypothesis, we are accepting the fact that there is a possibility that we are wrong.

For instance, with a 10% significance level, we are rejecting the null hypothesis when the statistical probability is less than 10%. However, at the 10% significance level, we know that the probability that a statistic lies outside of this range due to random sample error is 1 in 10. That is, that 1 in 10 times the statistic will fall below the critical value due to randomness alone. Likewise, at the 10% significance level, we know that the probability that a statistic from the same population lies inside of this range due to random sample error is 9 in 10. That is, that 9 in 10 times the statistic from the same population will fall above the significance level due to randomness alone. Thus, at most, with a 10% significance level, the null hypothesis will be correct only 90% of the time even if the samples come from the same population.

We can never definitively state that the null hypothesis is correct or incorrect. At most, we can state that we accept or reject it, but we are never really sure. And when we embrace or eschew the alternative hypothesis, we are also never really sure if it is right or wrong either.

Type I and Type II Errors

Each time we accept or reject the null hypothesis, we are running the risk of making an error. We could be accepting it when it is false, and we could be rejecting it when it is true.

Type I

Type I errors result from rejecting the null hypothesis when it is true.

For example, suppose we obtained the statistical probability of similarity to be 3% (p-value = 0.03) and were using the 5% significance level ($\alpha = 0.05$) for our hypothesis testing. We would reject the null hypothesis.

However, at this significance level, we know that 5% of the time the statistical probability of similarity would be less than 5% due to random sample error. That is, we expect to find p-values less than alpha at least 5% of the time anyway.

Who is to say that the obtained p-value = 3% < α = 10% isn't really just due to random sample error as opposed to being a real difference?

We can't be sure. We can only suspect.

While we would reject the null hypothesis in this case, we must acknowledge that we could be rejecting it when, in fact, it was true.

Rejecting the null hypothesis when the null hypothesis was indeed true is called a Type I error.

Type II

Type II errors arise from accepting the null hypothesis when it is false.

For example, suppose we obtained the statistical probability of similarity to be 30% (p-value = 0.30) and were using the 5% significance level ($\alpha = 0.05$) for our hypothesis testing. We would accept the null hypothesis.

However, being 30% similar isn't the same as being the same. There is still a chance that the differences are real.

Who is to say that the obtained p-value = 30% > α =5% isn't really due to real differences between the samples as opposed to being a random sample variation?

We can't be sure. We can only suspect.

While we would accept the null hypothesis in this case, we must acknowledge that we could be accepting it when in fact, it was false.

Accepting the null hypothesis when the null hypothesis was indeed false is called a Type II error.

Tradeoffs in Errors

We can increase our significance level to reduce the likelihood of rejecting the null hypothesis when it is true (Type I error), but, when we do, we increase the likelihood of accepting the null hypothesis when it is false (Type II error).

Alternatively, we can reduce our significance level to reduce the likelihood of accepting the null hypothesis when it is false (Type II error), but, when we do, we increase the likelihood of rejecting the null hypothesis when it is true (Type I error).

In either case, we are trading off the probability of making one type of error for another.

Decrease the significance level (smaller α)	Decrease the likelihood of Type I error Increase the likelihood of Type II error
Increase the significance level (larger α)	Increase the likelihood of Type I error Decrease the likelihood of Type II error

The only method of reducing the likelihood of making an error is to measure the entire population, rather than simply a sample. However, such a measure is prohibitively costly. As such, we accept that our conclusions are possibly wrong, but at least by measuring the items, we can state that our conclusions are accurate at a chosen significance level.

Sample Size

One method of reducing the chances of an error is to increase the sample size. However, random error is approximately proportional to one over the square root of the number of measures.

$$Random\ Error \propto \frac{1}{\sqrt{N}}$$

To reduce the random error by a factor of 2, we have to increase the sample size by a factor of 4. And to reduce the random error by a factor of 10, we have to increase the sample size by a factor of 100. Quickly, we realize that we must accept that our determinations may be imperfect, but perfection is too costly.

We accept that we must trade off cost for accuracy. At least, with the 5% significance level, we know we are right 95% of the time.